Prentice-Hall Foundations of Modern Organic Chemistry Series

KENNETH L. RINEHART, JR., Editor

umes published or in preparation

. ALLINGER and J. ALLINGER	**STRUCTURES OF ORGANIC MOLECULES**
TRAHANOVSKY	**FUNCTIONAL GROUPS IN ORGANIC COMPOUNDS**
STEWART	**THE INVESTIGATION OF ORGANIC REACTIONS**
SAUNDERS	**IONIC ALIPHATIC REACTIONS**
GUTSCHE	**CHEMISTRY OF CARBONYL COMPOUNDS**
PRYOR	**INTRODUCTION TO FREE RADICAL CHEMISTRY**
STOCK	**AROMATIC SUBSTITUTION REACTIONS**
RINEHART	**OXIDATION AND REDUCTION OF ORGANIC COMPOUNDS**
DePUY and CHAPMAN	**MOLECULAR REACTIONS AND PHOTOCHEMISTRY**
IRELAND	**ORGANIC SYNTHESIS**
DYER	**APPLICATIONS OF ABSORPTION SPECTROSCOPY OF ORGANIC COMPOUNDS**
BATES and SCHAEFER	**RESEARCH TECHNIQUES IN ORGANIC CHEMISTRY**
TAYLOR	**HETEROCYCLIC COMPOUNDS**
HILL	**COMPOUNDS OF NATURE**
BARKER	**ORGANIC CHEMISTRY OF BIOLOGICAL COMPOUNDS**
STILLE	**INDUSTRIAL ORGANIC CHEMISTRY**
SIM, RINEHART NORMAN, and GILBERT	**X-RAY CRYSTALLOGRAPHY, MASS SPECTROMETRY, AND ELECTRON SPIN RESONANCE OF ORGANIC COMPOUNDS**
BATTISTE	**NON-BENZENOID AROMATIC COMPOUNDS**

OXIDATION AND REDUCTION OF ORGANIC COMPOUNDS

Kenneth L. Rinehart, Jr.
Professor of Chemistry
University of Illinois

PRENTICE-HALL, INC., ENGLEWOOD CLIFFS, N.J.

Library of Congress Cataloging in Publication Data

Rinehart, Kenneth L.
Oxidation and reduction of organic compounds.

(Foundations of modern organic chemistry series)
Includes bibliographies.
1. Oxidation. 2. Reduction, Chemical
3. Chemistry, Organic. I. Title.
QD281.09R55 547'.23 72-1977
ISBN 0-13-647586-8
ISBN 0-13-647578-7 (pbk.)

To my parents

Printed in the United States of America

PRENTICE-HALL INTERNATIONAL, INC., London
PRENTICE-HALL OF AUSTRALIA, PTY. LTD., Sydney
PRENTICE-HALL OF CANADA, LTD., Toronto
PRENTICE-HALL OF INDIA PRIVATE LIMITED, New Delhi
PRENTICE-HALL OF JAPAN, INC., Tokyo

10 9 8 7 6 5 4 3 2 1

Contents

3

OXYGENATION OF HYDROCARBONS 37

4

REDUCTIVE CLEAVAGE IN OXYGEN-CONTAINING FUNCTIONAL GROUPS 59

5

OXIDATION OF ALCOHOLS AND PHENOLS 68

OXIDATION AND REDUCTION AT NITROGEN, SULFUR, AND PHOSPHORUS 121

Foreword

Organic chemistry today is a rapidly changing subject whose almost frenetic activity is attested by the countless research papers appearing in established and new journals and by the proliferation of monographs and reviews on all aspects of the field. This expansion of knowledge poses pedagogical problems; it is difficult for a single organic chemist to be cognizant of developments over the whole field and probably no one or pair of chemists can honestly claim expertise or even competence in all the important areas of the subject.

Yet the same rapid expansion of knowledge—in theoretical organic chemistry, in stereochemistry, in reaction mechanisms, in complex organic structures, in the application of physical methods—provides a remarkable opportunity for the teacher of organic chemistry to present the subject as it really is, an active field of research in which new answers are currently being sought and found.

To take advantage of recent developments in organic chemistry and to provide an authoritative treatment of the subject at an undergraduate level, the *Foundations of Modern Organic Chemistry Series* has been established. The series consists of a number of short, authoritative books, each written at an elementary level but in depth by an organic chemistry teacher active in research and familiar with the subject of the volume. Most of the authors have published research papers in the fields on which they are writing. The books will present the topics according to current knowledge of the field, and individual volumes will be revised as often as necessary to take account of subsequent developments.

The basic organization of the series is according to reaction type, rather than along the more classical lines of compound class. The first ten volumes in the series constitute a core of the material covered in nearly every one-year organic chemistry course. Of these ten, the first three are a general introduction to organic chemistry and provide a background for the next six, which deal with specific types of reactions and may be covered in any order. Each of the reaction types is presented from an elementary viewpoint, but in a depth not possible in conventional textbooks. The teacher can decide how much of a volume to cover. The tenth examines the problem of organic synthesis, employing and tying together the reactions previously studied.

The remaining volumes provide for the enormous flexibility of the series. These cover topics which are important to students of organic chemistry and are sometimes treated in the first organic course, sometimes in an intermediate course. Some teachers will wish to cover a number of these books in the one-year course; others will wish to assign some of them as outside reading; a complete intermediate organic course could be based on the eight "topics" texts taken together.

The series approach to undergraduate organic chemistry offers then the considerable advantage of an authoritative treatment by teachers active in research, of frequent revision of the most active areas, of a treatment in depth of the most fundamental material, and of nearly complete flexibility in choice of topics to be covered. Individually the volumes of the Foundations of Modern Organic Chemistry provide introductions in depth to basic areas of organic chemistry; together they comprise a contemporary survey of organic chemistry at an undergraduate level.

Kenneth L. Rinehart, Jr.
University of Illinois

Preface

This small volume on oxidation and reduction is the last to be written of those books in the Foundation of Modern Organic Chemistry Series which deal with types of reactions. It thus joins other volumes: *Ionic Aliphatic Reactions*, by Saunders, *The Chemistry of Carbonyl Compounds*, by Gutsche, *Introduction to Theoretical Chemistry*, by Pryor, *Aromatic Substitution Reactions*, by Stock, and *Molecular Reactions and Photochemistry*, by DePuy and Chapman. *Oxidation and Reduction of Organic Compounds* differs from those five books in its organization in that it lays less stress on mechanism and more on functional groups, in particular on methods of converting one functional group to another. The intent of this stress on the interconversion of groups is to prepare the student for *Organic Synthesis*, by Ireland, the logical next volume in the series, in much the same way that the volume *Organic Reaction Mechanisms*, by Stewart was designed to provide students with a transition to reactions from the volumes in the series dealing with compounds' structures (*Structures of Organic Molecules*, by the Allingers) and properties (*Functional Groups in Organic Compounds*, by Trahanovsky).

Oxidation and reduction play an important role in most synthetic sequences. Groups used to introduce other groups must then be oxidized or reduced to the level required in the products, or even removed entirely. Ring systems may be available in one oxidation state for synthesis but be needed in another oxidation state in the final product.

One reason for a reduced emphasis on mechanisms in the present volume is that the mechanisms of oxidations and reductions have been less studied than those of many other types of reactions. These mechanisms are discussed where they are known, but most are not known, even though many oxidation and reduction reactions are quite important. Moreover, those mechanisms that are understood fall into a number of mechanistic categories: some are ionic reactions; some are radical reactions; some involve carbonyl reductions; some are aromatic reactions. Inevitably, then, there is some overlap in the present volume with other volumes dealing with reactions in this series. Such overlap is indicated where the mechanisms are understood. Extensive additional references to the types of reactions discussed are provided at the ends of the chapters.

Throughout *Oxidation and Reduction*, stress has been laid on examples applicable to laboratory scale reactions which often involve expensive reagents rather than on commercial preparations on a large scale, where cheaper reagents can be employed. Large scale reactions are left to the book in this series by Stille, *Industrial Organic Chemistry*. Usually in carrying out an oxidation or reduction on a specific compound the conditions are important, and sometimes they are critical. Most of the reactions shown in *Oxidation and Reduction of Organic Compounds* can be documented, either as tested laboratory procedures such as organic syntheses, or from references to the research literature.

KENNETH L. RINEHART, JR.

1

Introduction

DEFINITIONS

Oxidation and reduction are familiar words used to describe a very large number of reactions, although the terms may be somewhat difficult to define precisely and comprehensively for organic chemistry. One traditional inorganic definition, involving a change in the valence of an element, is inapplicable, since the valence of carbon is nearly always 4. A second inorganic definition, involving the gain or loss of electrons, is difficultly applicable, since organic oxidations or reductions are not often carried out at electrodes and are usually irreversible.

One useful definition of oxidation for an organic compound is the gain of oxygen or loss of hydrogen; conversely, reduction of an organic compound is the loss of oxygen or gain of hydrogen. Thus, hydrogenation, the addition of hydrogen to a molecule, is an example of a reduction, whereas dehydrogeneration is an example of an oxidation. This definition covers most but not all oxidations and reductions of organic compounds. It can be ambiguous, since both hydrogen and oxygen are sometimes gained—or lost—from an organic compound in a single reaction. For example, water can be added to alkenes or eliminated from alcohols, and alkenes can be converted to glycols.

Although a completely comprehensive definition may be difficult, one can nearly always decide whether a compound is oxidized or reduced (or neither) in a particular reaction. One of the most useful criteria is the fate of the inorganic reactant. For example, in one conversion of an olefin to an α-glycol, osmium(VIII) is converted to osmium(VI); since the inorganic reagent is reduced, the organic compound must have been oxidized.

$$\mathrm{>C{=}C<} + \underset{Os^{VIII}}{OsO_4} \longrightarrow \underset{Os^{VI}}{\text{(cyclic osmate ester: } >C(-O)-C(-O)< \text{ with } Os(=O)_2)} \xrightarrow{H_2O} \underset{Os^{VI}}{>C(OH)-C(OH)<} + OsO_3$$

On the other hand, the addition of water to an alkene does not involve a change in the oxidation state of the oxygen atom, and this reaction is neither an oxidation nor a reduction.

$$\begin{array}{c} \diagdown\ \ \diagup \\ C \\ \| \\ C \\ \diagup\ \ \diagdown \end{array} + H_2O \longrightarrow \begin{array}{l} \diagdown\ \ \diagup \\ C-OH \\ | \\ C-H \\ \diagup\ \ \diagdown \end{array}$$

Thus, another useful definition might be that an organic compound is oxidized if the second reactant is reduced, and an organic compound is reduced if the second reactant is oxidized.

The present book describes only reactions that are generally regarded as oxidations or as reductions. In particular, those reactions involving only gain or loss of water and other acids or bases—hydrogen chloride or trimethylamine, for example—are not covered. These are found in another book in this series.* Since the reactions involved in oxidations and reductions are of diverse mechanistic types, the chapters are arranged according to the classifications of the compounds involved, beginning with the least oxygenated—hydrocarbons—and proceeding to the most oxygenated—acids. Nitrogen- and sulfur-containing compounds present somewhat different problems and are treated separately in a later chapter.

BALANCING OXIDATION-REDUCTION REACTIONS

In discussing oxidations and reductions of organic compounds, we shall usually not make the effort required to balance the equations; instead, we shall emphasize the nature of the organic reactants and products. Balancing oxidations or reductions can be difficult because these reactions often give several different products. Nevertheless, it is useful to know how to balance oxidation-reduction equations as an indication of the oxidation and reduction states involved, and also as a measure of the efficiency of the reagents employed.

In balancing an equation for an oxidation or a reduction, it is first necessary to know what all the products are. With this knowledge, the stoichiometry of the reaction may be obvious, and a balanced equation can be written from inspection. The hydrogenation of benzene to cyclohexane and the oxidation of benzyl alcohol to benzaldehyde are two reactions that can be balanced by inspection. Use of 3 moles of hydrogen balances the former, whereas the latter is balanced by adding 1 mole of water to the products.

Unbalanced:

$$\underset{(C_6H_6)}{\text{benzene}} + H_2 \xrightarrow[\text{catalyst}]{\text{Pt}} \underset{(C_6H_{12})}{\text{cyclohexane}}$$

* W. H. Saunders, *Ionic Aliphatic Reactions*, in Foundations of Modern Organic Chemistry Series, Prentice-Hall, Inc., Englewood Cliffs, N.J., 1965.

Balanced:

$$C_6H_6 + 3H_2 \xrightarrow{Pt} C_6H_{12}$$

Unbalanced:

$$C_6H_5{-}CH_2OH + MnO_2 \longrightarrow C_6H_5{-}CHO + MnO$$

Balanced:

$$C_6H_5{-}CH_2OH + MnO_2 \longrightarrow C_6H_5{-}CHO + MnO + H_2O$$

Sometimes, however, the stoichiometry of a reaction may not be obvious. In this case, as in inorganic chemistry, it is often best to divide the overall reaction into half-reactions, one for the species oxidized and the other for the species reduced. Water (H_2O), hydrogen ion (H^+), and hydroxide ion (OH^-) are used as needed to balance the two half-reactions. The half-reactions will not balance completely, however, without certain bookkeeping devices. Two such devices are common: one uses formal hydrogen and oxygen atoms, [H] and [O]; the other uses electrons (e^-). At the end the two half-reactions are combined to eliminate [H] and [O] or e^-. This usually requires multiplying one or both the half-reactions by a suitable factor. The two methods will now be illustrated for two common organic reactions.

Example 1. The reduction of nitrobenzene to aniline can be carried out with iron powder in hydrochloric acid; the reactants and products are those shown.

$$\underset{\text{nitrobenzene}}{C_6H_5{-}NO_2} + Fe \longrightarrow \underset{\text{aniline}}{C_6H_5{-}NH_2} + Fe^{3+}$$

The two unbalanced half-reactions are

$$C_6H_5{-}NO_2 \longrightarrow C_6H_5{-}NH_2 \qquad \text{(A)}$$

and

$$Fe \longrightarrow Fe^{3+} \qquad \text{(B)}$$

Balancing the half-reactions using [H] and [O] yields

$$C_6H_5\text{—}NO_2 + 6[H] \longrightarrow C_6H_5\text{—}NH_2 + 2H_2O \qquad (A1)$$

or

$$C_6H_5\text{—}NO_2 + H_2O \longrightarrow C_6H_5\text{—}NH_2 + 3[O] \qquad (A2)$$

and

$$Fe + 3H^+ + \tfrac{3}{2}[O] \longrightarrow Fe^{3+} + \tfrac{3}{2}H_2O \qquad (B1)$$

or

$$Fe + 3H^+ \longrightarrow Fe^{3+} + 3[H] \qquad (B2)$$

Note that the charges have been balanced on the two sides of each equation using hydrogen ion since the reaction is carried out in acid, and that [H] and [O] have been added to balance hydrogen or oxygen atoms removed or added. By comparing Equations (A1) and (A2) or (B1) and (B2), we see that one [O] is equivalent to two [H]'s. When we add half-reactions, [H]'s cancel one another by being on opposite sides of the equations; [H] and [O] cancel one another on the same side of the equations by giving water. To give the overall balanced equation, we combine two half-reaction equations. In this case it is more convenient to combine (A1) with (B2), but only after multiplying (B2) by 2. Note that the final equation balances, as it must.

$$C_6H_5\text{—}NO_2 + 6[H] \longrightarrow C_6H_5\text{—}NH_2 + 2H_2O \qquad (A1)$$

$$2 \times (Fe + 3H^+ \longrightarrow Fe^{3+} + 3[H]) \qquad (B2)$$

$$C_6H_5\text{—}NO_2 + 2Fe + 6H^+ \longrightarrow C_6H_5\text{—}NH_2 + 2Fe^{3+} + 2H_2O$$

To balance the equation using e^-, we add the electrons to balance excess charge on one side of the equation, and the half-reactions become

$$C_6H_5\text{—}NO_2 + 6H^+ + 6e^- \longrightarrow C_6H_5\text{—}NH_2 + 2H_2O \qquad (A3)$$

and

$$Fe \longrightarrow Fe^{3+} + 3e^- \qquad (B3)$$

Equation (B3) is multiplied by 2 and combined with (A3) to give the overall equation, which again must balance.

$$C_6H_5{-}NO_2 + 6H^+ + 6e^- \longrightarrow C_6H_5{-}NH_2 + 2H_2O \quad (A3)$$

$$2 \times (Fe \longrightarrow Fe^{3+} + 3e^-) \quad (B3)$$

$$C_6H_5{-}NO_2 + 2Fe + 6H^+ \longrightarrow C_6H_5{-}NH_2 + 2Fe^{3+} + 2H_2O$$

The latter method (involving e^-) is somewhat easier here; of course, the results are the same. Organic reactions, like inorganic reactions, can be regarded as 1-, 2-, or n-electron reductions or oxidations. The reduction of nitrobenzene to aniline is clearly a 6-electron reduction. By comparing Equations (A1) and (A3), we see that one [H] is equivalent to one e^-.

Example 2. The oxidation of an alcohol to the salt of an acid by potassium permanganate is carried out in aqueous base, the other product being manganese dioxide.

$$C_6H_5{-}CH_2OH + MnO_4^- \longrightarrow C_6H_5{-}COO^- + MnO_2$$

benzyl alcohol benzoate ion

We can balance this equation by half-reactions, as before, employing both the [H] and [OH] method, and the e^- method. Using [H] and [OH] to balance half-reactions gives those shown:

$$C_6H_5{-}CH_2OH + OH^- \longrightarrow C_6H_5{-}COO^- + 4[H] \quad (C1)$$

or

$$C_6H_5{-}CH_2OH + OH^- + 2[O] \longrightarrow C_6H_5{-}COO^- + 2H_2O \quad (C2)$$

and

$$MnO_4^- + \tfrac{1}{2}H_2O \longrightarrow MnO_2 + OH^- + \tfrac{3}{2}[O] \quad (D1)$$

or

$$MnO_4^- + 3[H] \longrightarrow MnO_2 + OH^- + H_2O \quad (D2)$$

To balance the overall equation, (C1) is multiplied by 3 and (D2) by 4.

$$3 \times \left(C_6H_5\text{-}CH_2OH + OH^- \longrightarrow C_6H_5\text{-}COO^- + 4[H] \right) \qquad (C1)$$

$$4 \times (MnO_4^- + 3[H] \longrightarrow MnO_2 + OH^- + H_2O) \qquad (D2)$$

$$3\,C_6H_5\text{-}CH_2OH + 4MnO_4^- \longrightarrow 3\,C_6H_5\text{-}COO^- + 4MnO_2 + OH^- + 4H_2O$$

Using e^- to balance the two half-reactions gives the following equations:

$$C_6H_5\text{-}CH_2OH + 5OH^- \longrightarrow C_6H_5\text{-}COO^- + 4H_2O + 4e^- \qquad (C3)$$

$$MnO_4^- + 2H_2O + 3e^- \longrightarrow MnO_2 + 4OH^- \qquad (D3)$$

Again, equations (C3) and (D3) must be multiplied by 3 and 4, respectively, to achieve an overall balanced equation.

$$3 \times \left(C_6H_5\text{-}CH_2OH + 5OH^- \longrightarrow C_6H_5\text{-}COO^- + 4H_2O + 4e^- \right)$$

$$4 \times (MnO_4^- + 2H_2O + 3e^- \longrightarrow MnO_2 + 4OH^-) \qquad (D3)$$

$$3\,C_6H_5\text{-}CH_2OH + 4MnO_4^- \longrightarrow 3\,C_6H_5\text{-}COO^- + 4MnO_2 + OH^- + 4H_2O$$

Of course, the balanced equation comes out the same, but here the [H][OH] method seems easier.

REFERENCES

*Augustine, R. L., ed., *Oxidation*, Marcel Dekker, Inc., New York, 1969.

*Augustine, R. L., ed., *Reduction*, Marcel Dekker, Inc., New York, 1968.

*Stewart, R., *Oxidation Mechanisms: Applications to Organic Chemistry*, W. A. Benjamin, Inc., New York, 1964.

*Waters, W. A., *Mechanisms of Oxidation of Organic Compounds*, Methuen & Co., Ltd., London, 1964.

*Wiberg, K. B., ed., *Oxidation in Organic Chemistry*, Part A, Academic Press, Inc., New York, 1965.

* References marked with asterisks will be cited in later chapters by author only, as Stewart, Waters, Wiberg, Augustine (*Reduction*), and Augustine (*Oxidation*).

PROBLEMS

1. In each of the following reactions, the organic reactants and products and the inorganic oxidizing and reducing agents are given. Using water, hydrogen ion, or hydroxide ion, complete and balance the equations.

(a) 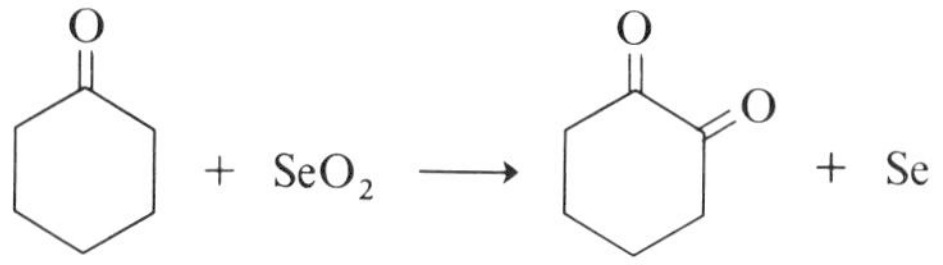 cyclohexanone + SeO_2 ⟶ cyclohexane-1,2-dione + Se

(b) 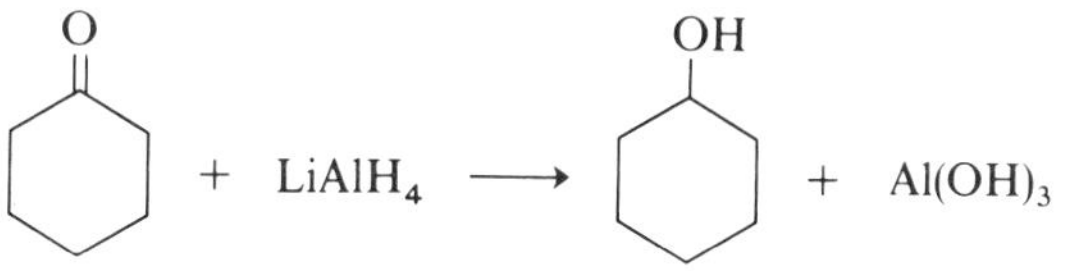 cyclohexanone + $LiAlH_4$ ⟶ cyclohexanol + $Al(OH)_3$

(c) 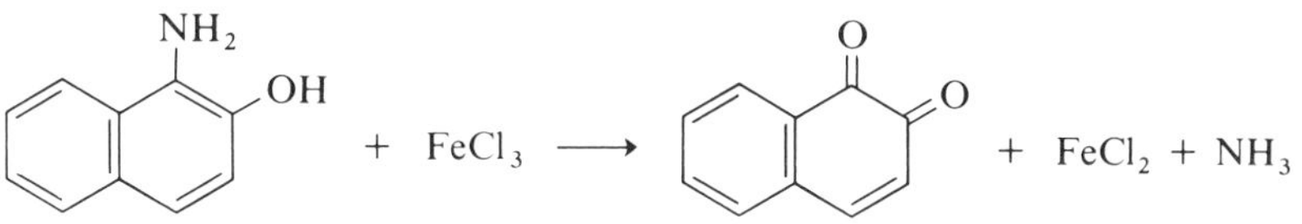 1-amino-2-naphthol + $FeCl_3$ ⟶ 1,2-naphthoquinone + $FeCl_2$ + NH_3

(d) $HOCH_2CH(OH)CH_2OH + IO_4^- \longrightarrow 2H_2C{=}O + HCOOH + IO_3^-$

(e) 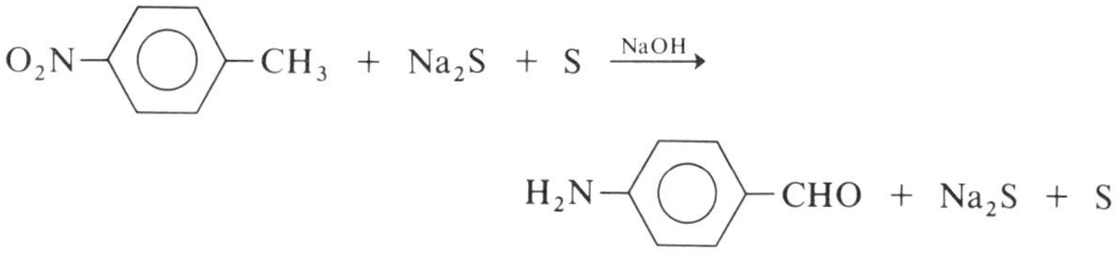 $O_2N{-}C_6H_4{-}CH_3 + Na_2S + S \xrightarrow{NaOH} H_2N{-}C_6H_4{-}CHO + Na_2S + S$

(f) $CS_2 + Cl_2 \longrightarrow CSCl_4 + S_2Cl_2$

(g) $CSCl_4 + Sn \xrightarrow{HCl} CSCl_2 + SnCl_4$

2. Are the organic reactants being oxidized or reduced or neither in the two equations shown?

(a) 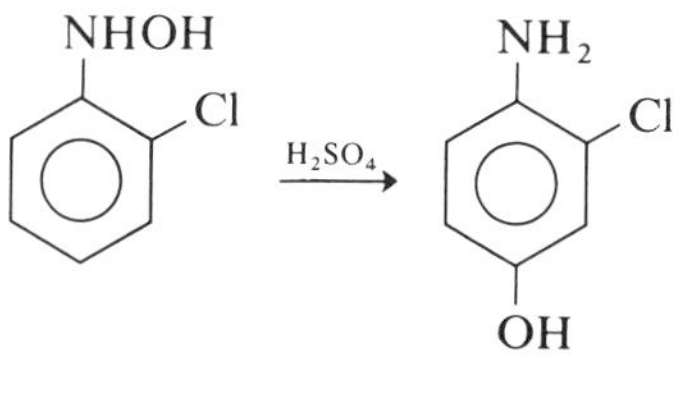 N-(2-chlorophenyl)hydroxylamine (NHOH, Cl) $\xrightarrow{H_2SO_4}$ 4-amino-3-chlorophenol (NH_2, Cl, OH)

(b) $(NC)_2C{=}C(CN)_2 \xrightarrow[\text{pyridine solution}]{H_2S,\ 0^\circ}$ 2,5-diamino-3,4-dicyanothiophene (NC, CN, H_2N, S, NH_2) + S

2

Hydrogenation and Dehydrogenation of Hydrocarbons

HYDROGENATION AND DEHYDROGENATION

Hydrocarbons can be arranged in a descending series of saturation according to the number of hydrogen atoms present per carbon atom: $C_nH_{2n+2}, C_nH_{2n}, C_nH_{2n-2}, \cdots$. By the definition found in Chapter 1, those compounds with the fewest hydrogen atoms are the most highly oxidized, and those with the most hydrogen atoms are the most highly reduced.

Formally, it should be possible to interconvert compounds in this series by removing or adding hydrogen by processes known as dehydrogenation and hydrogenation, respectively, which constitute examples of oxidation and reduction. In fact, some of these interconversions are relatively difficult and unusual, but others can be effected easily. The addition of hydrogen, a procedure called hydrogenation, is relatively easy to carry out in the laboratory and will be discussed first.

CATALYTIC HYDROGENATION OF ALKENES

The addition of hydrogen to a carbon-carbon double bond is a very important reaction that can be carried out on a small scale in a research laboratory and also on a large scale in an industrial plant. A catalyst is usually required to effect the reaction in either case; otherwise, hydrogenation is exceedingly slow. The catalyst is nearly always a metal or a compound that can be converted easily to a metal. The metal is not consumed in the reaction and is thus a true catalyst. By varying the type and amount of catalyst and such other variables as temperature and solvent, a large variety of different alkenes can be hydrogenated. The three examples shown include a compound containing an isolated carbon-carbon double bond, compounds containing alkene groups conjugated to benzene rings, and a compound with an alkene group conjugated to a carbonyl group. Although benzene rings and carbonyl groups can also be reduced by hydrogen, it is usually possible to hydrogenate an alkene group selectively.

$$\text{cyclohexene} + H_2 \xrightarrow[25^\circ]{5\%\ Pd/C} \text{cyclohexane}$$

1 atm pressure

$$C_6H_5{-}CH{=}CH_2 + H_2 \xrightarrow[20^\circ]{Ni(k)} C_6H_5{-}CH_2CH_3$$

3 atm

$$C_6H_5{-}CH{=}CH{-}C(=O){-}C_6H_5 + H_2 \xrightarrow[25^\circ,\ C_2H_5OAc\ solvent]{Pt} C_6H_5{-}CH_2CH_2C(=O){-}C_6H_5$$

3 atm

In the laboratory, the compounds to be hydrogenated are usually expensive, and the convenience of a procedure is also important. Therefore, the cost of the catalyst is not the decisive factor, and noble metals—palladium or platinum*—are usually chosen because they are the most active. These reactions are carried out, if possible, at atmospheric pressure and room temperature, where uptake of hydrogen can be measured accurately with a burette, as in the apparatus shown in Figure 2-1(a). This apparatus could have been used for the hydrogenation of cyclohexene, shown previously. Slightly higher pressures (to about 3 atm) are used in the Parr apparatus, shown in Figure 2-1(b), which employs carbonated beverage bottles as pressure vessels. The second and third hydrogenations, of styrene and chalcone (benzalacetophenone), could have been carried out in this apparatus.

Some hydrogenations require high pressures, either because the reduction to be effected is difficult or because the catalyst is relatively inactive. High-pressure hydrogenation is especially common in industrial operations, where cheap (and less active) catalysts are usually employed. Such hydrogenations are carried out in a pressure vessel (bomb), which can be heated and can be either stirred or rocked. A typical pressure apparatus of the rocking type, which can be heated to 350° and pressurized to 400 atm, is shown in Figure

* To provide the maximum surface area, one often prepares catalysts on supporting materials. In the first two equations shown, 5% Pd/C refers to 5% of palladium on a charcoal support, whereas Ni(k) refers to nickel supported on kieselguhr (a type of diatomaceous earth).

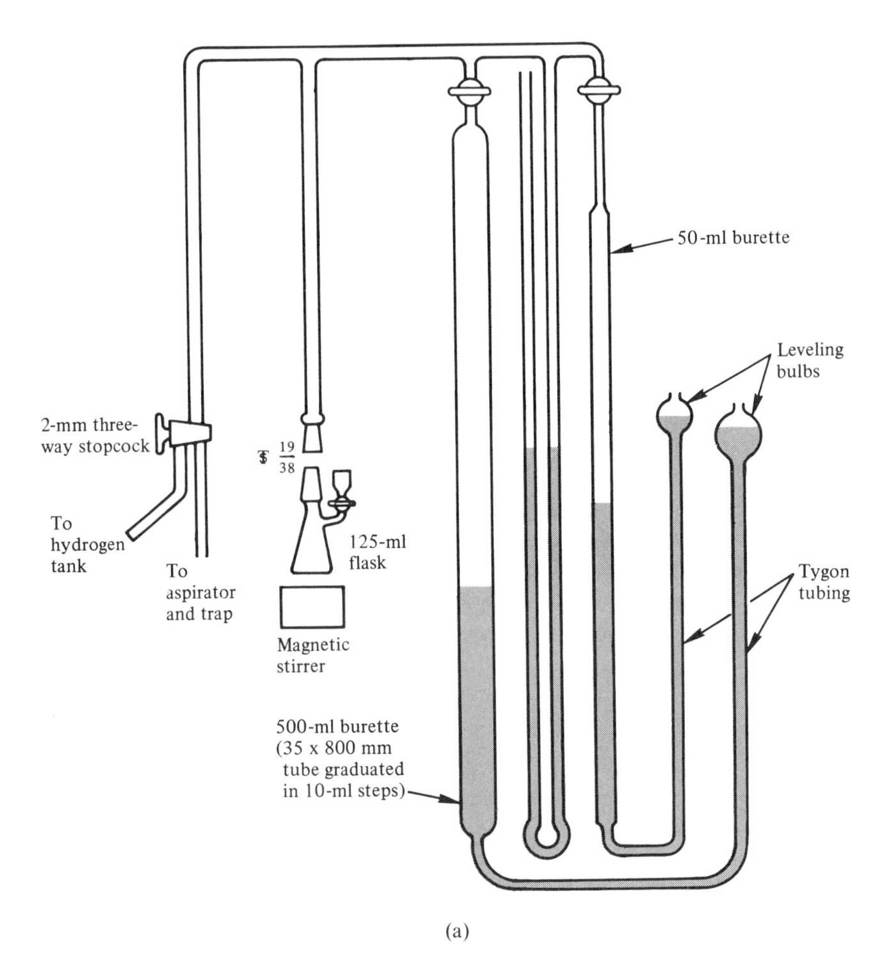

(a)

Fig. 2-1 Apparatus for catalytic hydrogenation. (a) Apparatus for quantitative atmospheric hydrogenation. The catalyst and solution of the sample to be reduced are stirred in the 125-ml flask. (From K. B. Wiberg, *Laboratory Technique in Organic Chemistry*, McGraw-Hill Book Company, Inc., New York, 1960, p. 228.) (b) Parr shaking apparatus for hydrogenation at 3–4 atm. The catalyst and solution of the sample to be hydrogenated are shaken in a carbonated beverage bottle. (c) High pressure hydrogenator for hydrogenation up to 6000 psi.

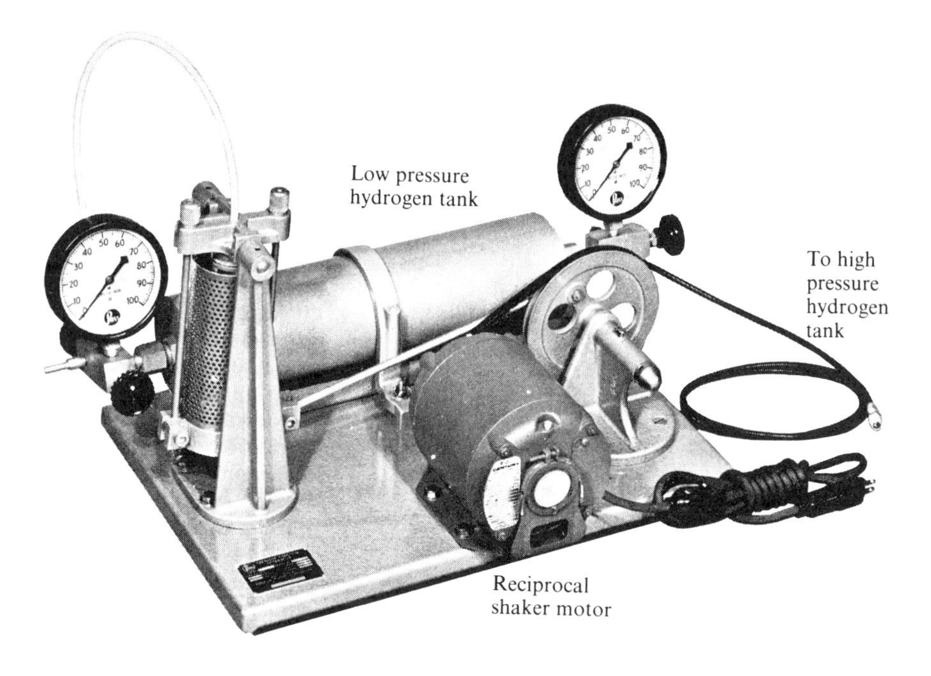

(b)

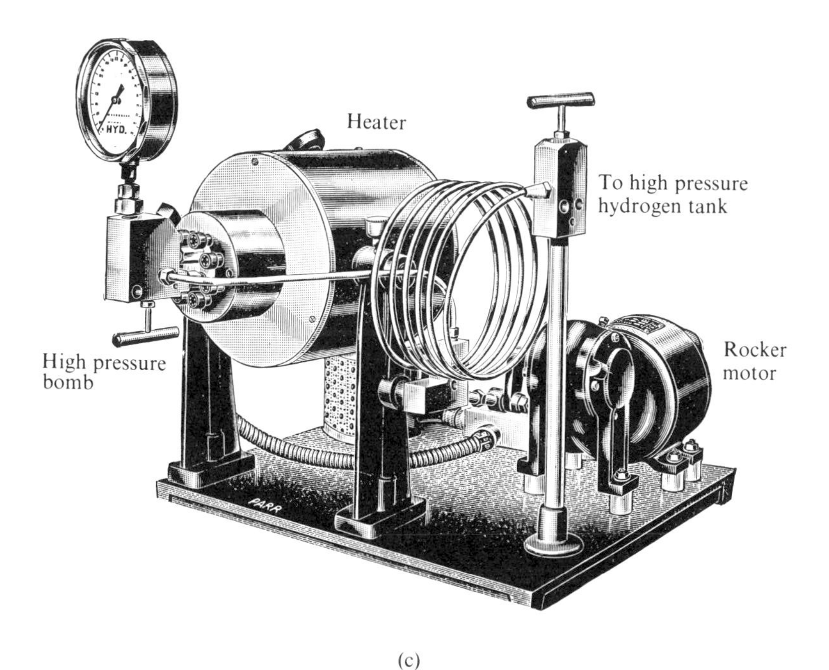

(c)

Fig. 2-1 (*contd.*)

2-1(c). It would be suitable for the high-pressure or high-temperature hydrogenations shown.

$$\text{cyclohexene} + H_2 \ (1\ \text{atm}) \xrightarrow[165^\circ]{CuCr_2O_4} \text{cyclohexane}$$

$$\text{cyclohexene} + H_2 \ (100\ \text{atm}) \xrightarrow[25^\circ]{Ni} \text{cyclohexane}$$

PREPARATION OF CATALYSTS

Since a catalyst is almost always required in hydrogenation, the nature of the catalyst plays a central role in the reaction. Great care must be taken in the preparation of catalysts; in particular, they must be finely divided and free of catalyst poisons, such as sulfur compounds. Palladium can be purchased as the free metal dispersed on carriers such as charcoal, barium sulfate, or calcium carbonate. Other catalysts are prepared in situ (in the reaction vessel), as shown in the following three equations: by reduction of platinum oxide to platinum (Adams catalyst); by reduction of palladium chloride to palladium (on a charcoal support); by dissolving aluminum into sodium hydroxide from a nickel-aluminum amalgam, leaving finely divided nickel (Raney nickel, available in several activities) behind.

$$PtO_2 + 2H_2 \longrightarrow Pt + 2H_2O$$

$$PdCl_2 + H_2 \longrightarrow Pd + 2HCl$$

$$\text{Ni-Al}_2 + 6NaOH \longrightarrow \text{Ni(R)} + 2Na_3AlO_3 + 3H_2$$

Industry uses some of the same catalysts employed in the laboratory, notably Raney nickel. However, many other less expensive catalysts are widely used, including nickel supported on kieselguhr and copper chromite. The following equations are not balanced because the products are complex.

$$NiNO_3 + H_2 \xrightarrow[\text{kieselguhr}]{450^\circ} \text{Ni(k)}$$

$$Cu(NO_3)_2 + (NH_4)_2CrO_4 \xrightarrow{400^\circ} \underset{\text{copper chromite}}{CuCr_2O_4}$$

To carry out a reduction using any of the five catalysts prepared thus far, we would introduce hydrogen from a high-pressure tank. It is also possible, however, to generate the hydrogen gas in situ for atmospheric pressure hydrogenations by reaction of sodium borohydride with acid. In this procedure, developed by H. C. Brown, the borohydride is first used to convert a platinum salt to the catalyst.

$$\underset{\text{excess}}{H_2PtCl_6 + NaBH_4} + 3C_2H_5OH \xrightarrow{25^\circ} Pt + B(OC_2H_5)_3 + NaCl + 5HCl + 2H_2$$

$$NaBH_4 + HCl + 3H_2O \longrightarrow H_3BO_3 + 4H_2 + NaCl$$

Another modification of hydrogenation techniques uses soluble catalysts. These behave differently from heterogeneous catalysts and have not yet achieved general use. However, they eliminate a major problem of using a heterogeneous system, viz., that the substance to be reduced must be first adsorbed onto a solid surface.

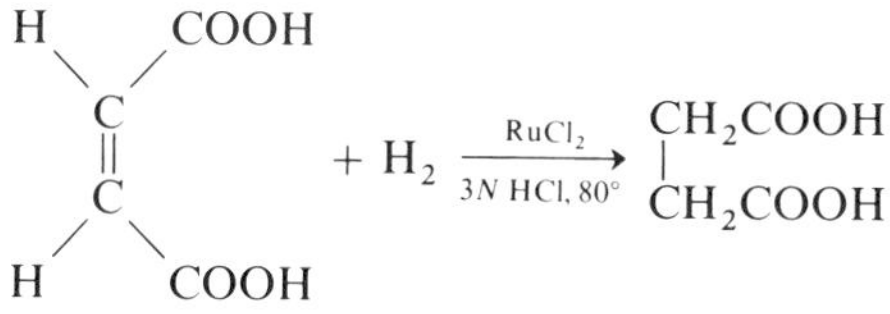

MECHANISM OF HYDROGENATION

The role of a catalyst in facilitating the addition of hydrogen is not completely clear, but apparently the free electrons on the metal surface serve to activate molecular hydrogen, converting it, at least in part, to a species that behaves like highly reactive atomic hydrogen, which then adds to the double bond. The weakening of the hydrogen-hydrogen bond in molecular hydrogen can be demonstrated by the interconversion of hydrogen and deuterium gases to a random mixture of H_2, D_2, and HD in the presence of a catalyst; in the absence of a catalyst, the mixture of H_2 and D_2 is quite stable. The catalyst also serves a second function. When the alkene is adsorbed by the catalyst, its π bond is weakened by partial bonding to the metal, so that addition of the "activated" hydrogen can occur.

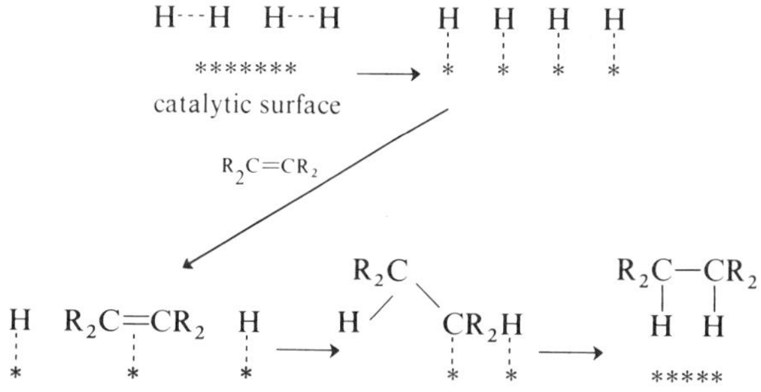

There is no general rule for the stereochemistry of hydrogenation, although cis addition often occurs; i.e., the two hydrogen atoms come on from the same side of the double bond. However, changes in solvent, hydrogen pressure, and catalyst can lead to predominant trans addition. Whereas a cis product would suggest the simultaneous addition of two hydrogen atoms to the double bond, isolation of a trans product indicates that the two hydrogen atoms must add separately. Another indication of step-by-step addition of hydrogen is provided by isomerization of the starting alkene to a different alkene, which sometimes occurs.

1,2-dimethylcyclohexene (CH_3, CH_3) + H_2 (1 atm) $\xrightarrow[\text{room temperature (r.t.)}]{Pd/Al_2O_3,\ HOAc}$ 25% cis (CH_3 CH_3) + 75% trans (CH_3, CH_3)

cis trans

1,2-dimethylcyclohexene (CH_3, CH_3) + H_2 (1 atm) $\xrightarrow[\text{r.t.}]{PtO_2,\ HOAc}$ 82% (CH_3 CH_3) + 18% (CH_3, CH_3)

$R_2C{=}C(R){-}C(H)\!<$ (H···*) $\longrightarrow$ $R{-}C(R)(H){-}C(R){=}C\!<$ (H···*)

For hydrogenation to occur, the alkene must approach the surface of the catalyst on which the hydrogen is adsorbed. Therefore, any interference with the approach to the catalyst surface will inhibit the reduction of the alkene. It is not surprising, then, that more highly substituted alkenes are generally reduced more slowly than their less highly substituted analogues, since substituents on the double bond hinder its approach to the catalyst surface. This selectivity can be demonstrated by the hydrogenation of limonene, in which it is possible to reduce the less hindered double bond selectively.

limonene + H_2 (3 atm) $\xrightarrow[60^\circ]{5\%\ Pt/C}$ 97%

limonene

The slow reduction of hindered alkenes is a problem of kinetics, having to do with the rate of reduction. On page 15, we shall discuss the thermodynamics of hydrogenation. Although it is often true that alkenes which are more slowly hydrogenated are also more thermodynamically stable, this does not have to be so, since the rate of hydrogenation is controlled by the free energy of activation for the hydrogenation reaction, whereas the heat of hydrogenation is associated with the enthalpy difference between the reactant and product.

CHEMICAL REDUCTION OF ALKENES

Catalytic hydrogenation is by far the most common method for the reduction of isolated carbon-carbon double bonds, and few other reagents can effect this reduction. However, recently discovered chemical procedures provide a useful complementary method. These involve reduction of the carbon-carbon double bond with diimide (also called diimine), HN=NH. This agent, which is consumed in the reaction, is effective in reductions because the second product formed, molecular nitrogen, is exceedingly stable, whereas diimide itself is not a stable compound and readily donates hydrogen with the formation of nitrogen. The transition state shown implies cis addition, which is found. Moreover, the absence of alkene isomerization indicates that the method is quite mild.

$$CH_2{=}CH(CH_2)_8COOH + HN{=}NH \longrightarrow CH_3(CH_2)_9COOH + N_2$$

$$\begin{array}{c} \diagdown\ \diagup \\ C \\ \| \\ C \\ \diagup\ \diagdown \end{array} + \begin{array}{c} H \\ \diagdown \\ N \\ \| \\ N \\ \diagup \\ H \end{array} \longrightarrow \begin{array}{ccc} C & \cdots H \cdots & N \\ \vdots\| & & \vdots\| \\ C & \cdots H \cdots & N \end{array} \longrightarrow \begin{array}{c} C{-}H \\ | \\ C{-}H \end{array} + \begin{array}{c} N \\ ||| \\ N \end{array}$$

Diimide is not a reagent that can be obtained from the laboratory shelf but must always be generated in situ. The methods shown are among those employed for this purpose.

$$H_2N{-}NH_2 + H_2O_2 \longrightarrow HN{=}NH + 2H_2O$$

$$2H_2N{-}OSO_3^- + 2OH^- \longrightarrow HN{=}NH + 2H_2O + 2SO_4^{2-}$$

$${}^-O_2C{-}N{=}N{-}CO_2^- + 2H^+ \longrightarrow HN{=}NH + 2CO_2$$

HEATS OF HYDROGENATION

The reaction of an alkene with hydrogen to given an alkane is exothermic; i.e., heat is given off. The amount of heat produced can be measured precisely in a calorimeter, where heat raises the temperature of some material (usually water) of known heat capacity. For most alkenes, the heat of hydrogenation is approximately 30 kcal/mole of the alkene. Although the heat evolved is the enthalpy of the reaction rather than the free energy,* its magnitude suggests that the equilibrium shown lies far to the right.

$$\text{alkene} + H_2 \rightleftharpoons \text{alkane} + \text{about 30 kcal/mole}$$

Although the value 30 kcal/mole is relatively constant, the small variations that occur are significant. In general, more highly substituted olefins are regarded as relatively stable, since they evolve less heat than their less highly substituted counterparts. Similarly, trans olefins are relatively stable because they evolve less heat than cis olefins. Heats of hydrogenation are an especially accurate reflection of the relative stabilities of isomeric alkenes when the alkenes are hydrogenated to give a common product, as in Figure 2-2.

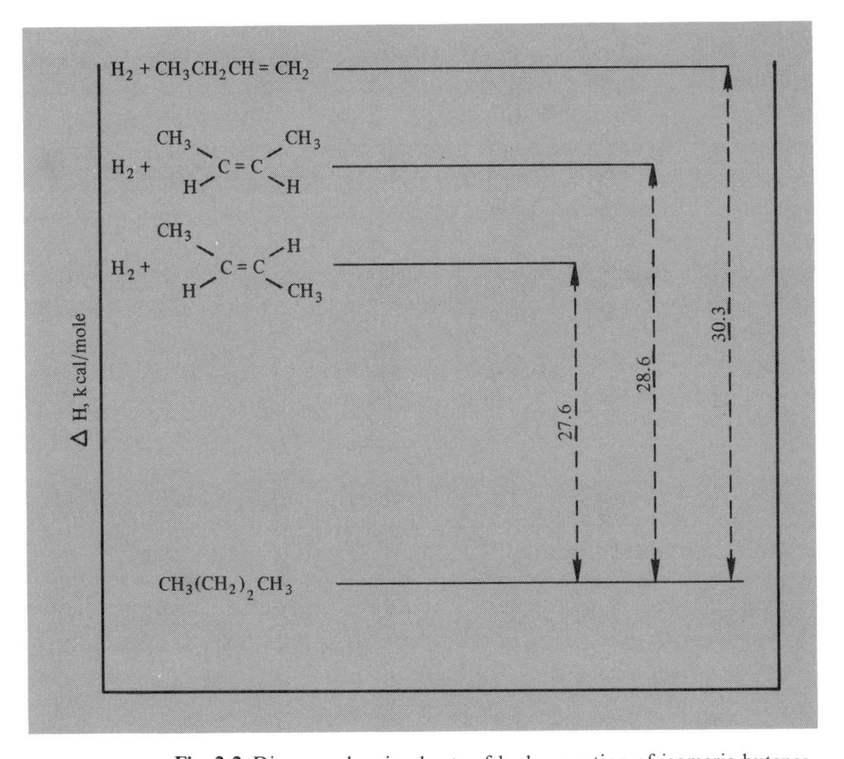

Fig. 2-2 Diagram showing heats of hydrogenation of isomeric butenes.

Heats of hydrogenation have also been used in evaluating strain energy in compounds containing small rings. The amounts of heat given off by different cycloalkenes are not quite reliable as an indication of the relative stabilities of the alkenes, since the products are different. Nevertheless, it is

* See, for example, R. Stewart, *Investigation of Organic Reactions*, in Foundations of Modern Organic Chemistry Series, Prentice-Hall, Inc., Englewood Cliffs, N. J., 1966.

clear from the values shown that cyclopentene, cyclohexene, cycloheptene, and cycloöctene are all relatively stable cycloalkenes, and that cyclopropene is the least stable (relative to the cycloalkane) of these simple compounds.

cyclopropene $+ H_2 \longrightarrow$ cyclopropane $+ 53.9$ kcal/mole

cyclobutene $+ H_2 \longrightarrow$ cyclobutane $+ 31.1$ kcal/mole

cyclopentene $+ H_2 \longrightarrow$ cyclopentane $+ 26.9$ kcal/mole

cyclohexene $+ H_2 \longrightarrow$ cyclohexane $+ 28.6$ kcal/mole

cycloheptene $+ H_2 \longrightarrow$ cycloheptane $+ 26.5$ kcal/mole

cycloöctene $+ H_2 \longrightarrow$ cycloöctane $+ 23.5$ kcal/mole

HYDROGENATION OF DIENES

Unconjugated dienes behave normally toward hydrogen, as they do toward most other reagents, consuming the reagent according to the reactivities of the isolated double bonds involved. We saw, for example, on page 14, that it is possible to reduce selectively the disubstituted double bond of limonene in the presence of its trisubstituted double bond.

Conjugated dienes also behave toward hydrogen as they do toward other reagents. Thus, 1,4-addition is observed, together with 1,2-addition.

$$CH_2{=}C(CH_3){-}CH{=}CH_2 + \underset{\text{1 mole}}{H_2} \xrightarrow{Pt} \underset{12\%}{CH_3CH(CH_3)CH{=}CH_2} + \underset{13\%}{CH_2{=}C(CH_3)CH_2CH_3}$$

$$+ \underset{15\%}{CH_3{-}C(CH_3){=}CHCH_3} + \underset{30\%}{(CH_3)_2CHC_2H_5} + \underset{30\%}{CH_2{=}C(CH_3){-}CH{=}CH_2}$$

When isomeric conjugated and unconjugated dienes are hydrogenated to the same alkane, the conjugated diene has a smaller heat of hydrogenation, indicating that it is somewhat more stable. In the example shown here, 1,3-pentadiene is about 6 kcal/mole more stable than 1,4-pentadiene.

$$CH_2{=}CHCH_2CH{=}CH_2 + 2H_2 \longrightarrow CH_3(CH_2)_3CH_3 + 60.6\ \text{kcal/mole}$$

$$CH_3CH{=}CHCH{=}CH_2 + 2H_2 \longrightarrow CH_3(CH_2)_3CH_3 + 54.1\ \text{kcal/mole}$$

HYDROGENATION OF ALKYNES

Carbon-carbon triple bonds react toward hydrogen very much like carbon-carbon double bonds, but add 2 moles of hydrogen. With care, the reaction can be stopped at the carbon-carbon double bond stage. This partial reaction is favored by the use of 1 mole of hydrogen and a small amount of catalyst. Special catalysts have been developed to assist in stopping the reaction after 1 mole of hydrogen. The most selective is palladium, specially prepared on barium sulfate or calcium carbonate and poisoned by quinoline. Hydrogenations with these catalysts (Lindlar catalysts) stop at the alkene, and the product is almost entirely the cis isomer. Of course, with more reactive catalysts, complete hydrogenation to the alkane is possible.

$$CH_3O_2C(CH_2)_3C{\equiv}C(CH_2)_3CO_2CH_3 + \underset{1\ \text{atm}}{H_2} \xrightarrow[CH_3OH,\ \text{r.t.}]{5\%\ Pd/BaSO_4,\ \text{quinoline}}$$

97% cis-$CH_3O_2C(CH_2)_3CH{=}CH(CH_2)_3CO_2CH_3$ (H and H on the same side of C=C)

CHEMICAL REDUCTION OF ALKYNES

It is possible to reduce the carbon-carbon triple bond by chemical means as well as by hydrogenation. Metals are usually employed, and these can give either cis or trans alkenes as products, depending upon which metal is used. As shown in the equations, sodium in ammonia gives the trans olefin, whereas zinc in acid gives the cis olefin. Both are consumed in the reactions.

$$CH_3(CH_2)_2C{\equiv}C(CH_2)_2CH_3 + 2Na + \underset{\text{liquid}}{2NH_3} \longrightarrow$$

trans-$CH_3(CH_2)_2CH{=}CH(CH_2)_2CH_3$ (H and $(CH_2)_2CH_3$ on one carbon; $CH_3(CH_2)_2$ and H on the other, H atoms trans) $+ 2NaNH_2$

$$C_6H_5C{\equiv}CC_6H_5 + Zn \xrightarrow[95^\circ,\ 24\ \text{hr}]{95\%\ HOAc}$$ cis-$C_6H_5CH{=}CHC_6H_5$ (H and H on the same side) $+ Zn(OAc)_2$

REDUCTION OF STRAINED CYCLIC HYDROCARBONS

The cyclopropane ring is abnormally strained (angle strain, Baeyer strain) and undergoes ring opening with hydrogen to give alkanes. Within a cyclopropane ring, any of the three ring carbon-carbon bonds can break, and a mixture of products is nearly always formed. For example, 1,2-di-*n*-octylcyclopropane gives a mixture of *n*-nonadecane and 9-methyloctadecane. Cyclopropane ring cleavage is relatively slow, and it is, therefore, possible to add hydrogen selectively to a cyclopropene carbon-carbon double bond, stopping at the cyclopropane stage. Cyclobutanes, and even cyclopentanes, react with hydrogen with ring opening, but these hydrogenations are considerably more difficult than hydrogenations of cyclopropanes, and higher temperatures or pressures must be used.

$$CH_3(CH_2)_7\text{-}(\triangle)\text{-}(CH_2)_7CH_3 + H_2 \xrightarrow[\text{HOAc, r.t.}]{PtO_2} CH_3(CH_2)_{17}CH_3 + CH_3(CH_2)_7CH(CH_3)(CH_2)_8CH_3$$

1 atm

↑ 1 mole H_2 at 1 atm, r.t., $Pd/BaSO_4$

$$CH_3(CH_2)_7\text{-(cyclopropene)-}(CH_2)_7CH_3$$

$$\triangle + H_2 \xrightarrow[80^\circ]{Ni} CH_3CH_2CH_3$$

$$\square + H_2 \xrightarrow[180^\circ]{Ni} CH_3(CH_2)_2CH_3$$

HYDROGENATION OF AROMATIC HYDROCARBONS

The addition of hydrogen to aromatic hydrocarbons is usually more difficult than the addition of hydrogen to alkenes, dienes, or alkynes, especially if relatively inactive catalysts are used. It is possible, for example, to add hydrogen over a nickel catalyst selectively to an alkene in the presence of a benzene ring. However, the aromatic ring will also consume hydrogen, which leads to formation of a cycloalkane. It is not possible to isolate an intermediate cyclic diene or alkene, since under the conditions required for hydrogenation of the aromatic system, these intermediate cyclic compounds would be hydrogenated rapidly. The preferred catalysts for hydrogenation of benzene rings are platinum, ruthenium, and rhodium. These often give mainly cis addition, but trans addition is also common.

C_6H_5–CH=CH$_2$ + H_2 (2.5 atm) —Ni(k), 20°→ C_6H_5–CH_2CH_3 —H_2, 100 atm, Ni(k), 125°→ cyclohexyl–H, CH_2CH_3

NH_2, $N(C_2H_5)_2$ + $3H_2$ (135 atm) —RuO_2, 100°, dioxane→ 75% NH_2, $N(C_2H_5)_2$

In condensed ring aromatic systems, one ring is often somewhat more easily hydrogenated than the others. For example, naphthalene can be reduced selectively to give tetralin, whereas anthracene and phenanthrene both form their 9,10-dihydro derivatives. In all cases, more vigorous conditions of reduction give total hydrogenation.

+ $2H_2$ (4 atm) —PtO_2, HOAc, 45°→ —$3H_2$(4 atm), Rh/C, C_2H_5OH, 45°→ H, H

+ H_2 (65 atm) —Ni(R), C_2H_5OH, 50°→

+ H_2 (65 atm) —Ni(R), C_2H_5OH, 50°→ —$6H_2$ (165 atm), Ni(R), C_2H_5OH, 160°→ H, H, H, H

HYDROGENATION OF HETER CLIC AROMATIC COMPOUNDS

Although this chapter de primarily with the hydrogenation and dehydrogenation of hydrocarb s, whereas Chapter 8 treats oxidation and reduction of ni n-contain compounds, the hydrogenation of heterocyclic aromatic compounds i ufficiently similar to that of aromatic hydrocarbons that it, too, can be ated effectively here as well. The examples shown demonstrate that the me catalysts can be employed and that, as with benzenoid compounds, ductions of nitrogen- and oxygen-containing aromatic compounds proceed give the completely hydrogenated products. (It is the pyridir nucleus tha s hydrogenated in preference to the benzene nucleus in quin ne.) Thiophe e and its derivatives are somewhat different, in that nickel t s to remove lfur from them, giving hydrocarbons. Like most sulfur-co ning compo ds, they poison almost all catalysts.

pyridine $+ 3H_2$ (4 atm) $\xrightarrow[\text{HOAc, 40°}]{PtO_2}$ piperidine

quinoline $+ . H_2$ (3 .tm) $\xrightarrow[\text{HOAc, r.t.}]{PtO_2}$ 1,2,3,4-tetrahydroquinoline

2,5-dimethylpyrrole $+ 2H_2$ (3 atm) $\xrightarrow[\text{HOAc, r.t.}]{5\%\ Rh/Al_2O_3}$ *cis*-2,5-dimethylpyrrolidine

furan $+ 2H_2$ (4 atm) $\xrightarrow[C_2H_5OH,\ 40°]{Ni(R)}$

thiophene $+ 4H_2$ (1 atm) $\xrightarrow[C_2H_5OH,\ 80°]{Ni(R)}$ $CH_3(CH_2)_2CH_3 + H_2S$

HEATS OF HYDROGENATION OF AROMATIC COMPOUNDS

The heat (enthalpy) evolved during the hydrogenation of an aromatic compound to its saturated counterpart (cycloalkane or heterocycle) is less than one would expect based on the number of double bonds for a canonical formula. This means that the aromatic compound itself must be of lower energy than a hypothetical cyclic triene or heterocyclic diene. The amount by which the heat of hydrogenation of an aromatic compound differs from that expected for the hypothetical diene or triene containing localized double bonds is called the resonance energy of the compound. For benzene, the heat of hydrogenation is 49.8 kcal/mole, whereas the hypothetical triene, cyclohexatriene, should evolve $3 \times 28.8 = 86.4$ kcal/mole; thus, the resonance energy of benzene is calculated to be $86.4 - 49.8 = 36.6$ kcal/mole. The enthalpy relationships of benzene, the hypothetical cyclohexatriene, and cyclohexane are summarized in Figure 2-3. This concept of resonance energy and its application to aromatic compounds is discussed in detail

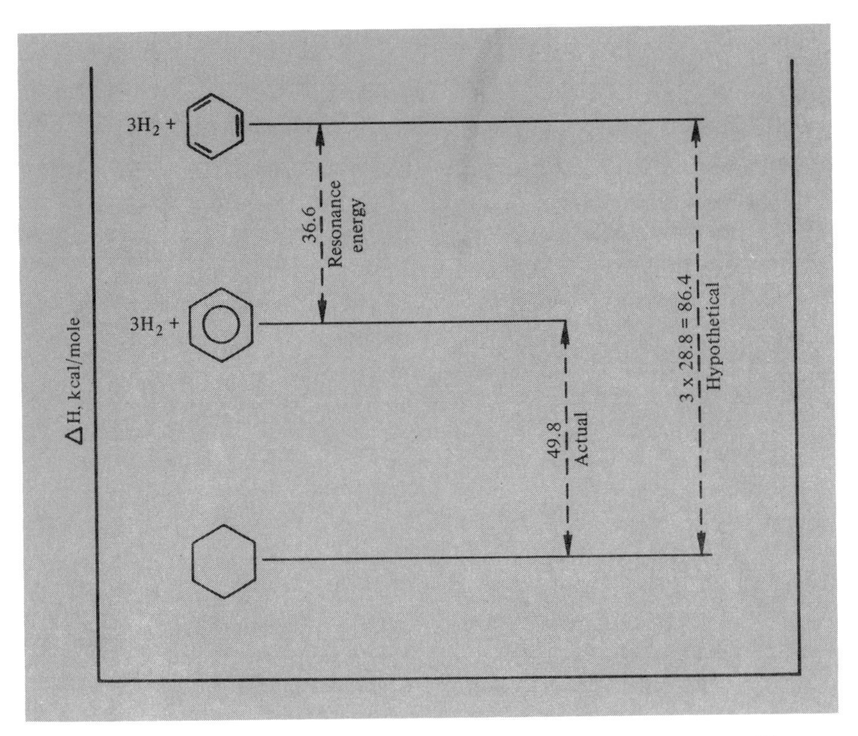

Fig. 2-3 Diagram showing the calculation of the resonance energy of benzene from the difference between the heats of hydrogenation of benzene and of a hypothetical cyclohexatriene.

elsewhere in this series.*

$$C_6H_6 + 3H_2 \longrightarrow C_6H_{12} + 49.8 \text{ kcal/mole}$$

Heats of hydrogenation (ΔH_h) have been used to assign resonance energies (R.E. values) to other cyclic compounds as well, as illustrated:

	$+3H_2$	$+4H_2$	$+5H_2$	$+7H_2$	$+2H_2$
ΔH_h (kcal)	72.8	92.6	99.0	130.8	36.6
R.E. (kcal)	6.7	13.2	28.3	28.0	17.2

CHEMICAL REDUCTION OF THE BENZENOID NUCLEUS

Like the chemical reduction of carbon-carbon double bonds, the chemical reduction of aromatic rings is rather difficult and not often employed. However, for many aromatic compounds, the Birch reduction can be used to advantage. The reducing agent in this reaction is an alkali metal (the metal, not the cation), which donates an electron to the aromatic compound, forming the alkali metal cation and a radical anion. The latter unstable species then abstracts a proton from a hydrogen donor to give a radical, which, in turn, reacts with a second alkali metal atom to give an anion. The second anion reacts, like the first, with a proton donor to give the product, a dihydroaromatic compound—usually an unconjugated cyclohexadiene. Ammonia or aliphatic amines function as solvents for the metal, whereas alcohols function as proton donors. Sometimes ethers are added to help dissolve the aromatic substrate.

$$C_6H_6 + Na \underset{}{\overset{\text{liq } NH_3}{\rightleftharpoons}} [C_6H_6]^{\cdot -} Na^+ \xrightarrow{C_2H_5OH} C_6H_7^{\cdot} + C_2H_5O^-$$

$$C_6H_7^{\cdot} \xrightarrow{Na} Na^+ \, C_6H_7^- \xrightarrow{C_2H_5OH} C_6H_8$$

* See, for example, L. Stock, *Chemistry of Aromatic Compounds*, in Foundations of Modern Organic Chemistry Series, Prentice-Hall, Inc., Englewood Cliffs, N. J., 1968.

Birch reduction does not work as well with compounds having aromatic rings substituted by electron donors or with those which are sterically hindered. Sometimes these can be reduced somewhat easier with lithium metal, as illustrated for anisole. An example of a particularly difficult reduction is the steroid derivative shown. In this reaction, it was necessary to employ an ether solvent and the tertiary alcohol as hydrogen donor. From these and the two following examples, we see that a strong directive effect is operative. Groups that are electron donors (such as methoxyl) end up on one of the remaining carbon-carbon double bonds, whereas electron-withdrawing groups, such as carboxyl, end up on one of the reduced carbon atoms. Despite difficulties, the Birch reduction can be very useful, since it gives a product (a dihydroaromatic compound) not obtainable by catalytic hydrogenation.

OCH_3 —[Li, liq NH_3; C_2H_5OH; $(C_2H_5)_2O$]→ OCH_3 (major product) + OCH_3 (minor product)

H_3C, OH, H, H, H, H, CH_3O —[Na, liq NH_3; t-C_4H_9OH; O]→ H_3C, OH, H, H, H, CH_3O

CO_2H —[Na, liq NH_3; C_2H_5OH]→ —[HCl; H_2O]→ CO_2H

$+ 2Na + 2C_2H_5OH$ —[liq NH_3; $(C_2H_5)_2O$]→ $+ 2NaOC_2H_5$

CATALYTIC DEHYDROGENATION OF ACYCLIC AND CYCLIC HYDROCARBONS

The conversion of relatively saturated hydrocarbons to unsaturated cyclic hydrocarbons is an exceedingly important process in industry. In the research laboratory, the dehydrogenation reaction is also useful, chiefly in structure proofs, where it allows the chemist to gain information on the nature of the ring system in an unknown compound. Hydrogen can be removed from the more saturated compound either catalytically or chemically.

Only the former method, catalytic dehydrogenation over a metal catalyst which is not consumed, is employed industrially,* where in the petrochemical industry fully saturated straight-chain hydrocarbons can be dehydrogenated to aromatic hydrocarbons. This provides a cheaper process than the isolation of aromatic hydrocarbons from coal tar, and most of the benzene and toluene produced now comes from the dehydrogenation route. Examples for the preparation of benzene and toluene from *n*-hexane and *n*-heptane are shown. These aromatic hydrocarbons can also be prepared from branched chain hydrocarbons containing 6 or 7 carbon atoms; since conditions are exceedingly strenuous, carbon-carbon bond making and breaking occur to give the most stable products, the aromatic compounds.

$$CH_3(CH_2)_4CH_3 \xrightarrow[35\ atm]{450°,\ Pt} \text{benzene} + 3H_2$$

$$CH_3(CH_2)_5CH_3 \xrightarrow[MoO_3-Al_2O_3]{500°,\ 16\ atm} \text{toluene } (C_6H_5CH_3) + 3H_2$$

Catalytic dehydrogenations in the laboratory are normally performed under less strenuous conditions than in industry, and fully saturated hydrocarbons can not often be employed. In general, at least one double bond should be present in the compound for dehydrogenation to occur. For example, methylcyclohexene readily gives toluene, and tetralin gives naphthalene. Even under these conditions, somewhat milder than those employed industrially, catalytic dehydrogenation is often accompanied by loss or rearrangement of alkyl groups, as illustrated in the example on the next page, which produces phenanthrene. It is thus not a perfect tool in structural elucidation but does often give some insight into the ring system.

$$\text{1-methylcyclohexene } (CH_3) \xrightarrow[350°]{Pt/C} \text{toluene } (CH_3) + 2H_2$$

$$\text{tetralin} \xrightarrow[\text{reflux, 4 hr; } CO_2 \text{ sweep}]{30\%\ Pd/C} \text{naphthalene} + 2H_2$$

* This process is discussed in greater detail in J. K. Stille, *Industrial Organic Chemistry*, in Foundations of Modern Organic Chemistry Series, Prentice-Hall, Inc., Englewood Cliffs, N. J., 1968.

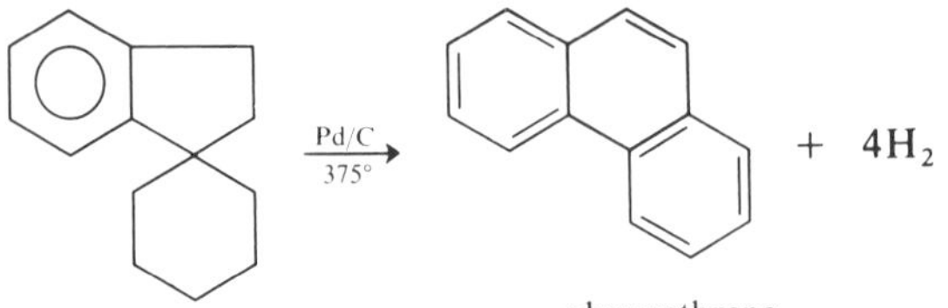

$$\xrightarrow[375^\circ]{Pd/C} \quad + \; 4H_2$$

phenanthrene

Catalytic dehydrogenations are the reverse of hydrogenations and are thus equilibria. Hydrogen gas must then be removed in order to shift the equilibrium forward to convert completely to the aromatic hydrocarbon; otherwise, a disproportionation occurs to give a mixture of a cycloalkane and the aromatic hydrocarbon. One can achieve the removal of hydrogen by passing the reactant over the catalyst. Once away from the catalyst, hydrogen, product, and starting material no longer react, and the excess starting material can be recycled. This procedure was that employed for the preparation of toluene, shown on page 25. As an alternative, one can add an easily reduced substance, such as indene, that cannot become fully aromatic and thus consumes hydrogen.

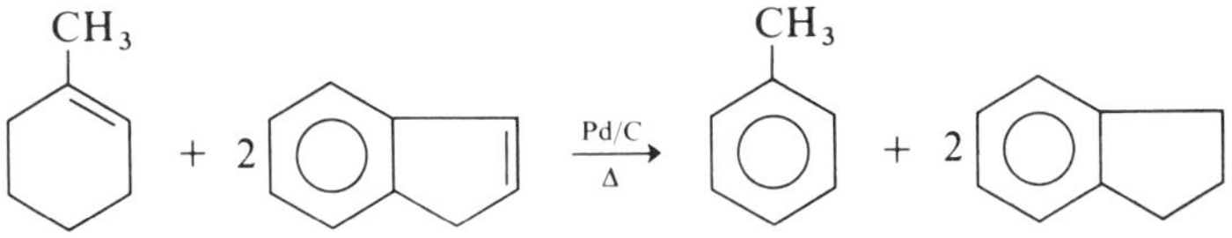

$$CH_3 \quad + \; 2 \quad \xrightarrow[\Delta]{Pd/C} \quad CH_3 \quad + \; 2$$

indene

CHEMICAL DEHYDROGENATION OF CYCLIC HYDROCARBONS

Many relatively saturated hydrocarbons can be converted by certain chemical reagents (as well as by catalytic dehydrogenation) to aromatic hydrocarbons. Oxygen is too strong an oxidant, since it gives largely combustion, together with oxygenated products; but the next two elements of this group in the periodic table—sulfur and selenium—are also oxidizing agents and are very effective in removing hydrogen. The overall reaction involves much the same reactants and products as catalytic dehydrogenation, but sulfur and selenium dehydrogenations can be carried out under somewhat milder conditions. At least one double bond is nearly always required in the substrate except at very high temperatures, as shown for the dehydrogenation of decalin. Of the two, sulfur is the more reactive element and requires a lower temperature.

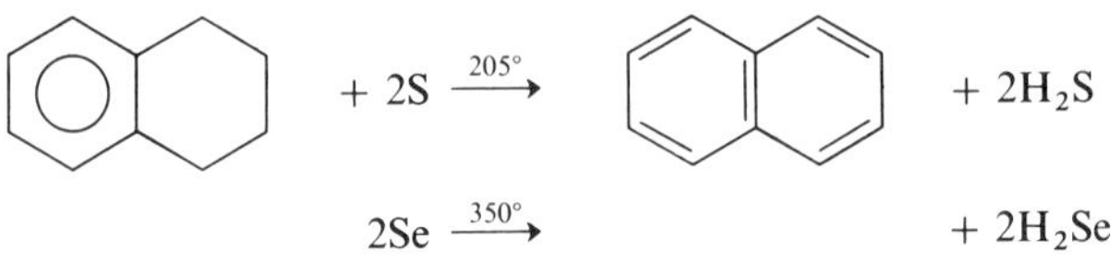

$$+ \; 2S \xrightarrow{205^\circ} \quad + \; 2H_2S$$

$$2Se \xrightarrow{350^\circ} \quad + \; 2H_2Se$$

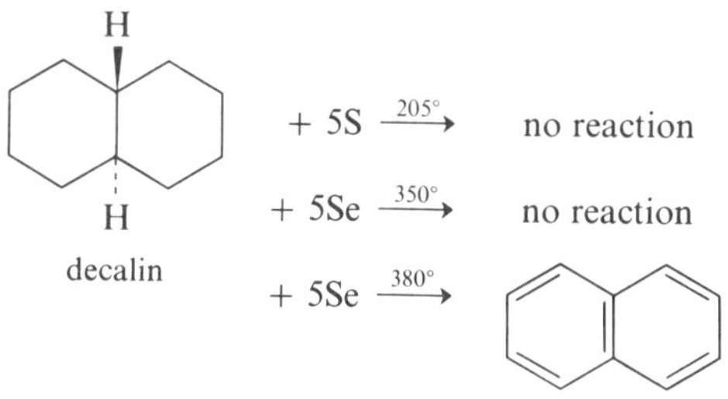

H

H

decalin

+ 5S $\xrightarrow{205°}$ no reaction

+ 5Se $\xrightarrow{350°}$ no reaction

+ 5Se $\xrightarrow{380°}$

Somewhat fewer rearrangements occur with sulfur and selenium dehydrogenation than with catalytic dehydrogenation, although alkyl groups can migrate or be lost and oxygen atoms can also be lost.

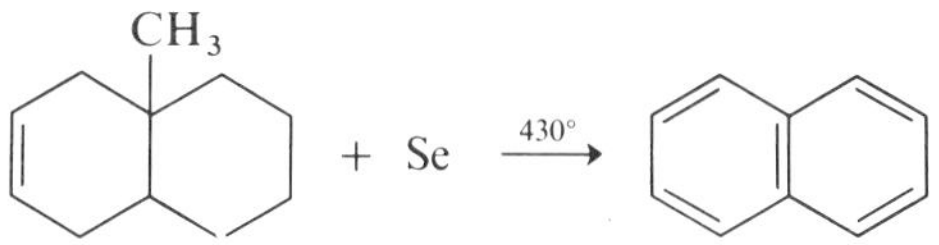

CH_3

+ Se $\xrightarrow{430°}$

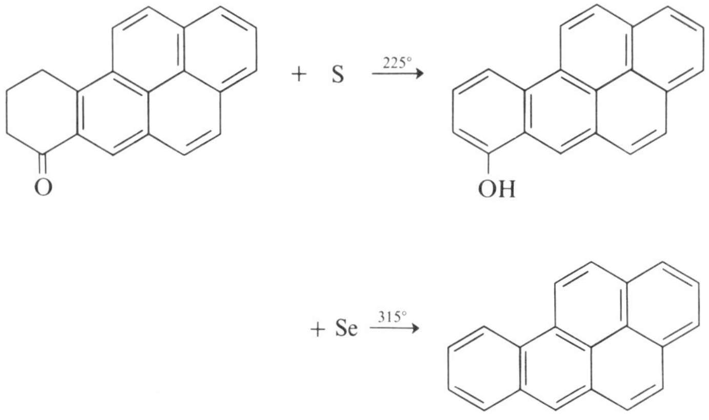

O + S $\xrightarrow{225°}$ OH

+ Se $\xrightarrow{315°}$

Chemical dehydrogenation, like catalytic dehydrogenation, can be useful as a tool in establishing structures of unknown compounds. An example of the use of dehydrogenation in structure proof is found in the conversion of the terpene cadinene to the naphthalene derivative, cadalene. Elemental analyses and molecular weight determinations had showed cadinene to have the molecular formula $C_{15}H_{24}$. When treated with sulfur, it gave cadalene ($C_{15}H_{18}$), a known compound, and was therefore deduced to be a hexahydro derivative of cadalene. The positions of the double bonds were subsequently established by oxidations. (However, cadalene is also formed from the monocyclic compound zingiberene, so dehydrogenation is by no means an infallible guide.)

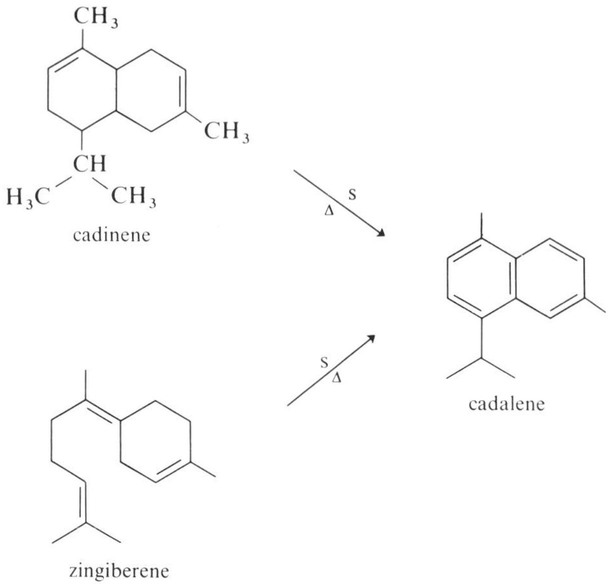

CH_3

CH

H_3C CH_3

cadinene

S

Δ

cadalene

S

Δ

zingiberene

Much milder reagents, especially quinones, are available for dehydrogenation when the hydroaromatic compounds are more nearly aromatic. An example is chloranil, tetrachloro-*p*-benzoquinone, which can be employed in dehydrogenating tetralin to naphthalene.

The conversion of quinones to hydroquinones is a reversible reaction and is discussed in more detail in Chapter 5. Some quinones that are particularly effective for dehydrogenation are chloranil, tetrachloro-*o*-benzoquinone, and 2,3-dichloro-5,6-dicyanobenzoquinone. All these quinones have electron-withdrawing substituents and are relatively more stable in the reduced form.

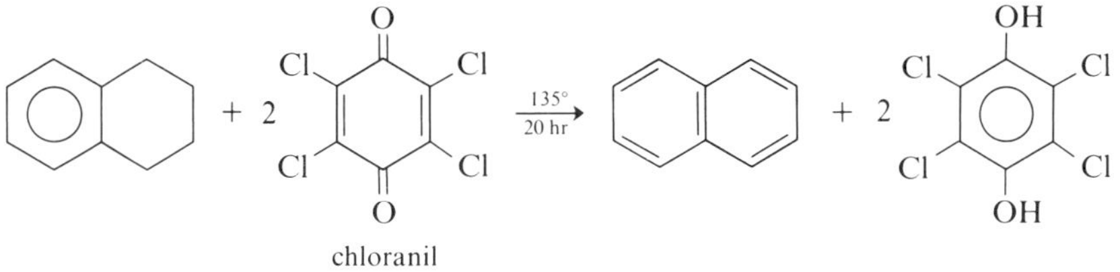

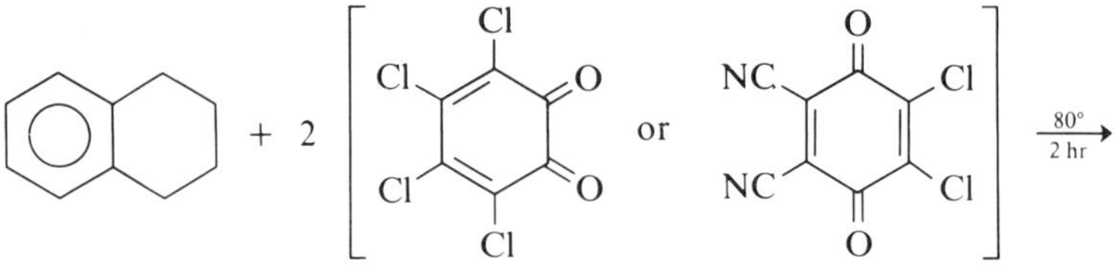

+ 2 [chloranil] $\xrightarrow[20\ hr]{135°}$ + 2

chloranil

+ 2 [or] $\xrightarrow[2\ hr]{80°}$

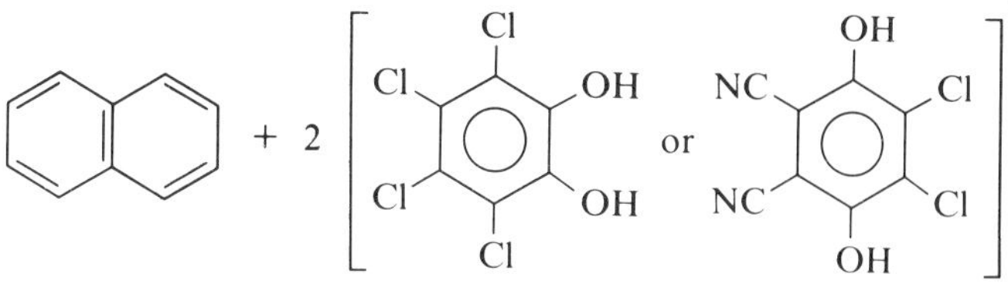

The mechanism of quinone dehydrogenation apparently involves hydride ion transfer from the cycloalkene to the quinone, followed by transfer of a proton.

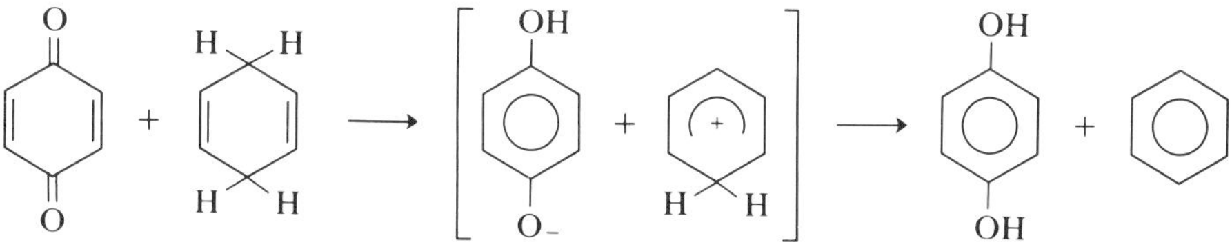

In all the dehydrogenation reactions shown, either catalytic or chemical, only aromatic compounds were obtained as products, not alkenes, dienes, and alkynes, because the more highly unsaturated compounds are even more reactive and are dehydrogenated more rapidly than their precursors—until a stable aromatic nucleus is obtained. Thus, dehydrogenation can go from saturated to aromatic compounds, and hydrogenation from aromatic to saturated compounds, but intermediates along the way between these two cyclic systems are difficult to isolate.

INTRODUCTION OF ALKENE BONDS INTO CYCLIC COMPOUNDS

An exception to the general rule that saturated compounds cannot be converted readily to alkenes is found in some compounds where a position is activated by a carbonyl group. In these compounds, it is often possible to introduce a carbon-carbon double bond adjacent (alpha, beta) to the carbonyl group, or in a position allylic to it (gamma, delta to an alpha, beta unsaturated carbonyl group) by treating the ketone with an oxidizing agent. This example of dehydrogenation has been especially useful in the chemistry of steroids, and a number of specialized techniques have been developed. Among the useful reagents for effecting this reaction are manganese dioxide, selenium dioxide,* and quinones such as chloranil. The apparent reason that chloranil does not convert the hydroaromatic compound completely

* Manganese dioxide is sometimes incorrectly regarded as being specific in oxidizing only allylic and benzylic alcohols (Chapter 5). Selenium dioxide is a powerful oxidizing agent and can effect a number of other sorts of reactions (Chapter 3).

to an aromatic compound in the case shown is that a blocking methyl group at the ring juncture prevents aromatization. Quinones are usually not sufficiently powerful as dehydrogenating agents to cause removal or rearrangement of alkyl groups.

$COCH_2OAc$ ---- OH, HO, H_3C, H_3C, O + Cl, Cl, Cl, Cl, O, O →

H_3C $COCH_2OAc$ ---- OH, H_3C, O + OH, Cl, Cl, Cl, Cl, OH

H_3C, H_3C, O $\xrightarrow[C_6H_6,\ reflux]{MnO_2}$ H_3C, H_3C, O

CH_3OOC CH_2COOCH_3 2 O + SeO_2 $\xrightarrow[reflux]{t\text{-}C_4H_9OH}$ CH_3OOC CH_2COOCH_3 2 O + Se + $2H_2O$

Mercuric acetate can also introduce a conjugated double bond. At high temperatures, this reagent is capable of dehydrogenating a saturated molecule, but it is difficult to predict where the unsaturation will be introduced, and the reaction is not very useful.

H_3C $C_{10}H_{19}$, H_3C, AcO $\xrightarrow[CHCl_3,\ HOAc]{Hg(OAc)_2,\ 25°}$ H_3C $C_{10}H_{19}$, H_3C, AcO

$$\xrightarrow[CCl_4,\ HOAc]{Hg(OAc)_2,\ 100^\circ}$$

COUPLING REACTIONS OF ACETYLENES

A formal dehydrogenation that is sometimes useful takes advantage of the peculiar nature of the acetylenic hydrogen atom, which is at the same time acidic and susceptible to oxidation. Mono-alkyl acetylenes react with mild oxidizing agents, notably oxygen in the presence of base and cuprous salts (where the actual oxidizing agent is presumed to be cupric ion), to give conjugated diacetylenes (1,3-diynes). An interesting example is found in the preparation of isomycomycin, an isomerization product of the antibiotic mycomycin.

$$2RC{\equiv}CH + O_2 \xrightarrow[NH_4OH]{Cu^+} RC{\equiv}C{-}C{\equiv}C{-}R$$

$$CH_3C{\equiv}C{-}C{\equiv}CH + HC{\equiv}C{-}CH{=}CH{-}CH{=}CHCH_2CN$$

$$\downarrow CH_3OH,\ H_2O,\ 5\ hr,\ r.t.;\ CuCl,\ O_2$$

$$CH_3C{\equiv}C{-}C{\equiv}C{-}C{\equiv}C{-}CH{=}CH{-}CH{=}CHCH_2CN$$

$$\downarrow CH_3OH,\ HCl,\ 55^\circ$$

$$\downarrow CH_3OH,\ KOH$$

$$CH_3(C{\equiv}C)_3(CH{=}CH)_2CH_2COOH$$

isomycomycin

$$\uparrow KOH,\ H_2O,\ r.t.$$

$$HC{\equiv}C{-}C{\equiv}C{-}CH{=}C{=}CH(CH{=}CH)_2CH_2COOH$$

mycomycin

The acetylene coupling reaction has also been used in the synthesis of cyclic polyunsaturated compounds containing 12 to 30 carbon atoms. Subsequent hydrogenation of these polyynes gives, under carefully controlled conditions, cyclic polyenes, and those with the appropriate number of conjugated carbon atoms have many of the characteristics of aromatic compounds.* In general, the acetylene coupling reaction has proved to be one of the most useful methods for the preparation of polyunsaturated cyclic hydrocarbons.

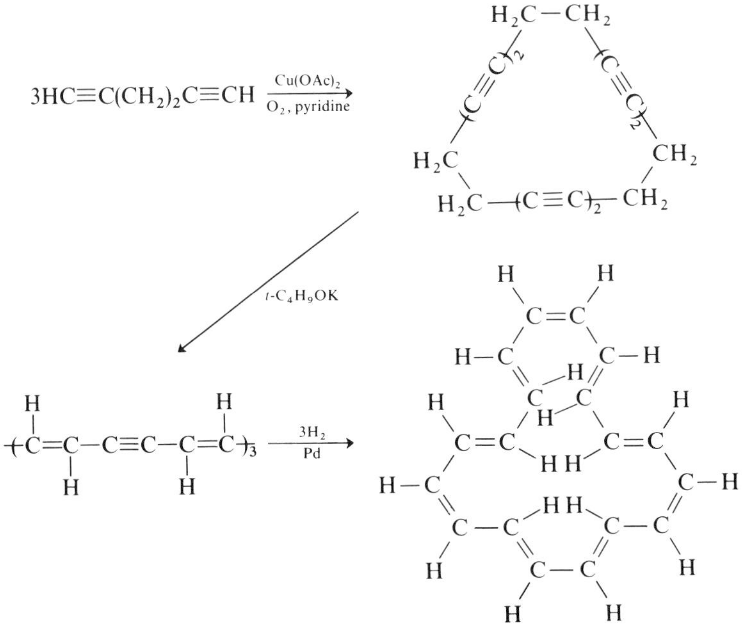

REFERENCES

Augustine, R. L., *Catalytic Hydrogenation*, Marcel Dekker, Inc., New York, 1965.

Augustine (*Reduction*): Chap. 2, M. Smith, "Dissolving Metal Reductions."

*Bentley, K. W., ed., *Elucidation of Structures by Physical and Chemical Methods*, Part *1*, Vol. 11 of *Technique of Organic Chemistry*, A. Weissberger, ed., John Wiley & Sons, Inc., New York, 1963, Chap. 9 (F. J. McQuillin, "Reduction and Hydrogenation in Structural Elucidation").

*Bentley (Part 1): Chap. 10, Z. Valenta, "Dehydrogenation."

Bond, G. C., and P. B. Wells, "The Mechanism of the Hydrogenation of Unsaturated Hydrocarbons on Transition Metal Catalysts," *Advan. Catal.*, **15**, 91 (1964).

* See, for example, L. Stock, *Chemistry of Aromatic Compounds*, in Foundations of Modern Organic Chemistry Series, Prentice-Hall, Inc., Englewood Cliffs, N. J., 1968, pp. 16–17.

Burwell, R. L., Jr., "The Mechanism of Heterogeneous Catalysis," *Chem. Eng. News*, **44**, 56 (August 22, 1956).

Burwell, R. L., Jr., "Deuterium as a Tracer in Reactions of Hydrocarbons on Metallic Catalysts," *Accounts Chem. Res.*, **2**, 289 (1969).

Foerst, W. A., ed., *Newer Methods of Preparative Organic Chemistry*, Vol. I, Interscience Publishers, New York, 1948: P. A. Plattner, "Dehydrogenation with Sulfur, Selenium, and Platinum Metals," pp. 21–60.

Freifelder, M., *Practical Catalytic Hydrogenation; Techniques and Applications*, John Wiley & Sons, Inc., New York, 1971.

Hansch, C., "The Dehydrocyclization Reaction," *Chem. Rev.*, **53**, 353 (1953).

*House, H. O., *Modern Synthetic Reactions*, 2nd ed., W. A. Benjamin, Inc., New York, 1972, Chap. 1 ("Catalytic Hydrogenation and Dehydrogenation") and Chap. 3 ("Dissolving Metal Reductions and Related Reactions").

Hückel, W., "Reduction of Hydrocarbons by Metals in Liquid Ammonia," *Fortschr. Chem. Forsch.*, **6**, 197 (1966).

Jackman, L. M., "Hydrogenation-Dehydrogenation Reactions," *Advan. Org. Chem.*, **2**, 329 (1960).

Newham, J., "Hydrogenolysis of Small Carbon Rings," *Chem. Rev.*, **63**, 123 (1963).

Rylander, P. N., *Catalytic Hydrogenation over Platinum Metals*, Parts 1 ("Catalysts, Equipment and Conditions"), 2 ("Hydrogenation of Carbon-Carbon Unsaturation"), and 5 ("Hydrogenation of Aromatics"), Academic Press, Inc., New York, 1967.

Siegel, S., "Stereochemistry and the Mechanism of Hydrogenation of Unsaturated Hydrocarbons," *Advan. Catal.*, **16**, 123 (1966).

Smith, H., "Hydrogenation of Unsaturated Systems with Metal-Ammonia Reagents," in *Organic Reactions in Liquid Ammonia*, John Wiley & Sons, Inc., New York, 1963, pp. 212–79.

PROBLEMS

1. Write detailed mechanisms for the following reactions, including all intermediates and/or transition states.

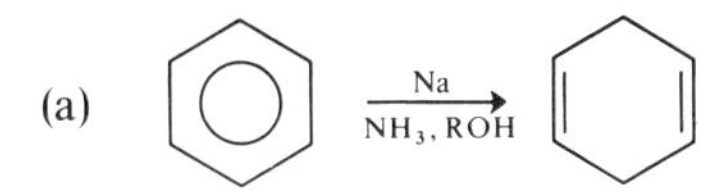

* References marked with asterisks will be cited in later chapters by author only, such as Bentley (Part 1 or Part 2), Foerst, House, etc.

(b) 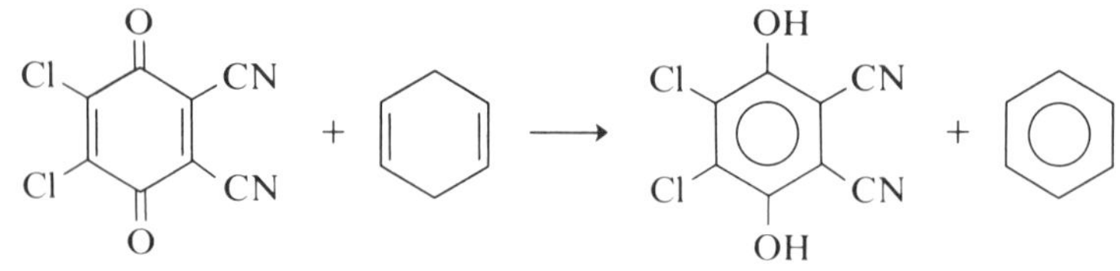 [2,3-dichloro-5,6-dicyano-1,4-benzoquinone] + [1,4-cyclohexadiene] $\longrightarrow$ [2,3-dichloro-5,6-dicyanohydroquinone] + [benzene]

(c) 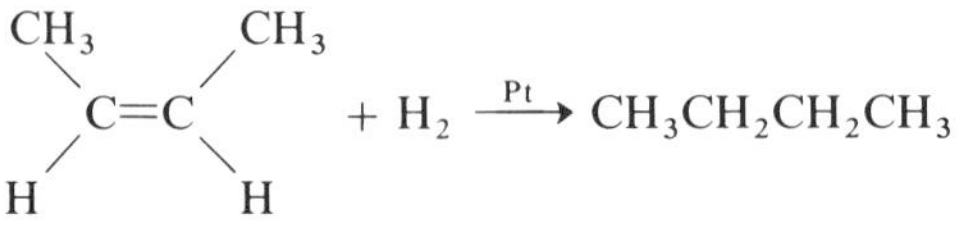 $CH_3(H)C{=}C(H)CH_3 + H_2 \xrightarrow{Pt} CH_3CH_2CH_2CH_3$

2. Suggest reaction conditions (reagents, catalysts, solvents, etc.) suitable for effecting the following conversions. More than one step may be required.

(a) 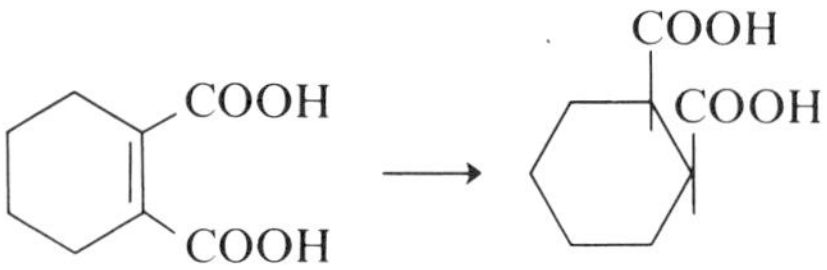 [cyclohexene-1,2-dicarboxylic acid (COOH, COOH)] $\longrightarrow$ [cis-cyclohexane-1,2-dicarboxylic acid (COOH, COOH)]

(b) 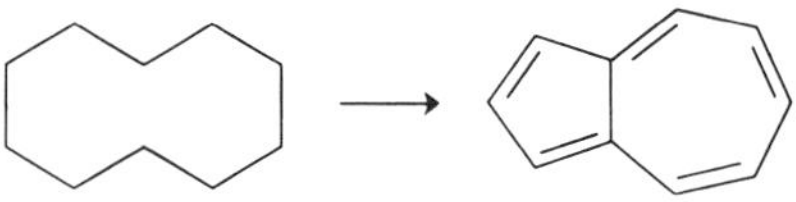 [cyclodecane] $\longrightarrow$ [azulene]

(c) 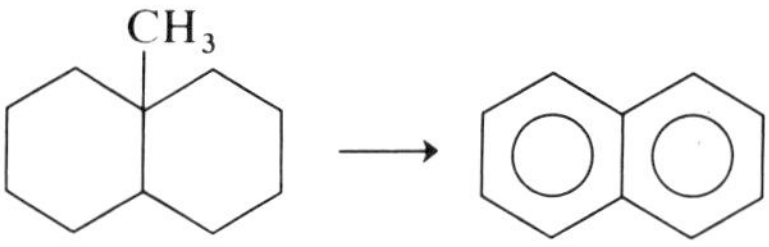 [9-methyldecalin (CH_3)] $\longrightarrow$ [naphthalene]

(d) 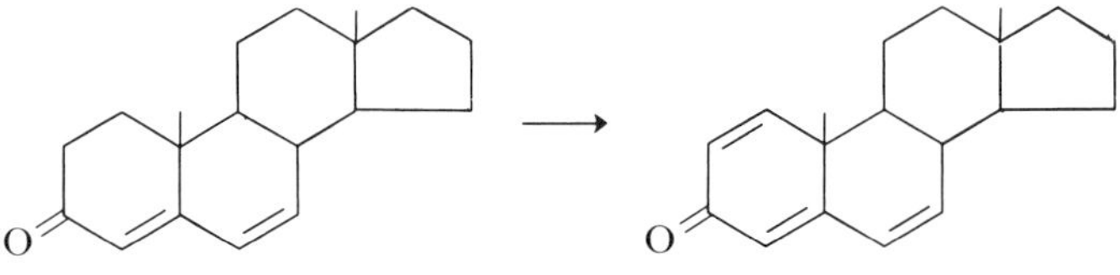 [steroid 4,6-dien-3-one (O)] $\longrightarrow$ [steroid 1,4,6-trien-3-one (O)]

(e)  $C_6H_5{-}COCH_3 \longrightarrow C_6H_{11}{-}COCH_3$

(f) 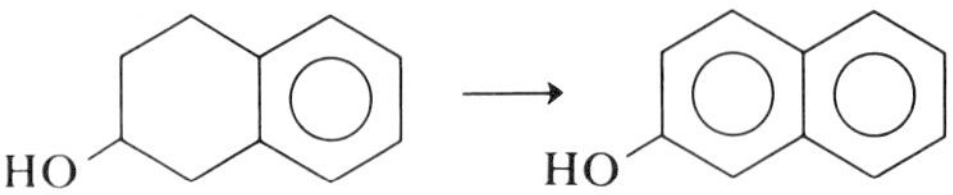 [HO-tetrahydronaphthalene] $\longrightarrow$ [HO-naphthalene]

(g) 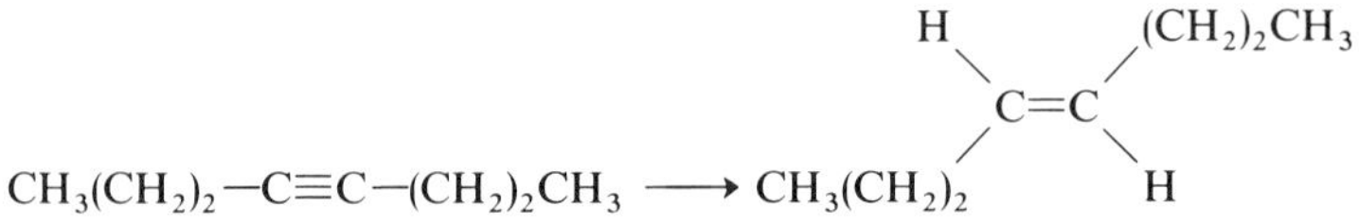 $CH_3(CH_2)_2{-}C{\equiv}C{-}(CH_2)_2CH_3 \longrightarrow CH_3(CH_2)_2(H)C{=}C(H)(CH_2)_2CH_3$ (trans)

(h) 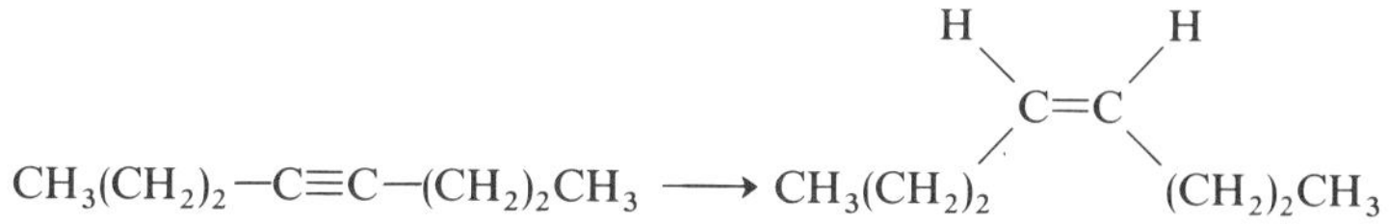 $CH_3(CH_2)_2-C\equiv C-(CH_2)_2CH_3 \longrightarrow$ H, H, C=C, $CH_3(CH_2)_2$, $(CH_2)_2CH_3$

(i) 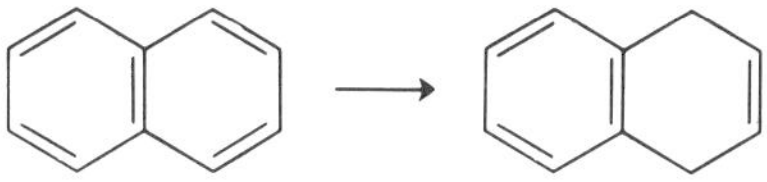 →

(j) 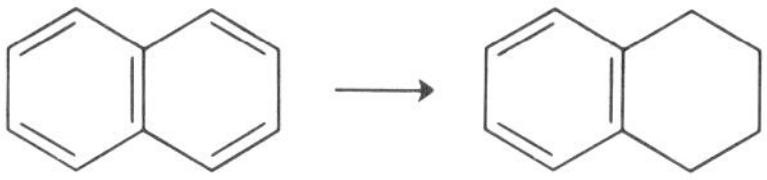 →

(k)  →

(l) 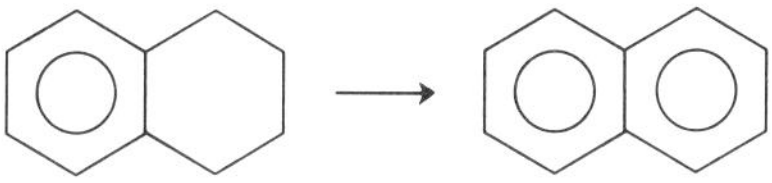 →

(m) 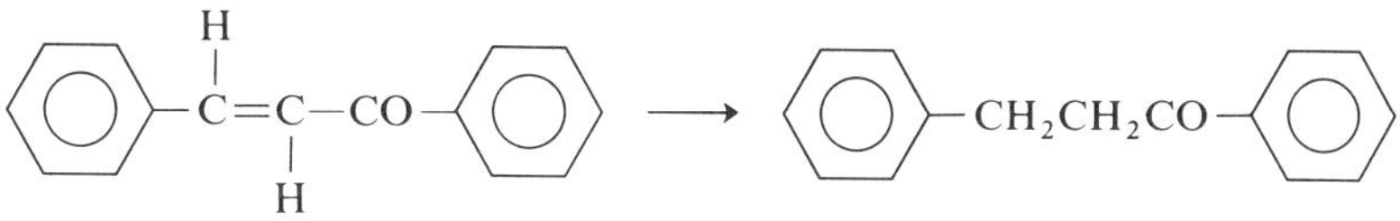 H, C=C—CO, H → CH_2CH_2CO

(n) 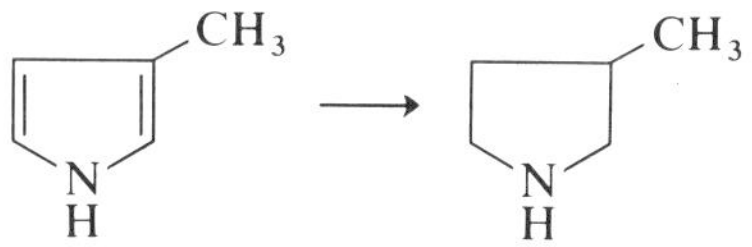 CH_3, N, H → CH_3, N, H

3. Show all the expected products of the following reactions, including their stereochemistry, where appropriate.

(a) 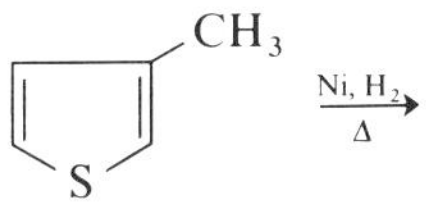 CH_3, S $\xrightarrow[\Delta]{Ni, H_2}$

(b) 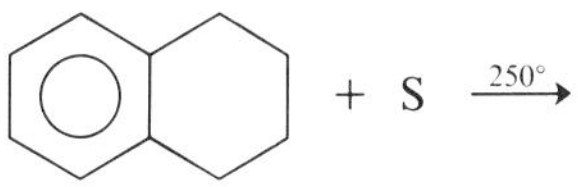 + S $\xrightarrow{250°}$

(c) 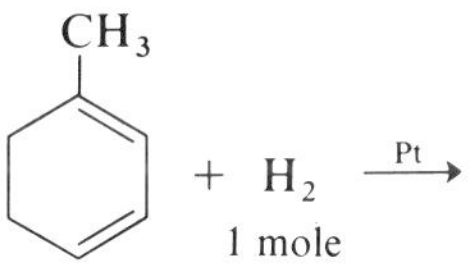 CH_3 + H_2 $\xrightarrow{Pt}$
1 mole

(d) 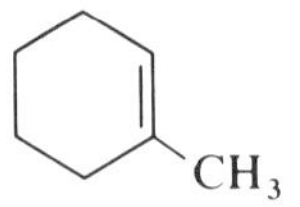 [1-methylcyclohexene] $+ H_2PtCl_6 + NaBH_4 + HCl \longrightarrow$
excess

(e) 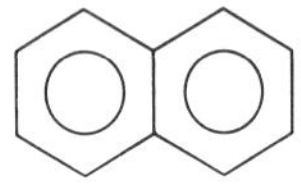 [naphthalene] $+ 5H_2 \xrightarrow[HOAc]{PtO_2}$

(f) $CH_3CH_2C{\equiv}CCH_2CH_3 \xrightarrow[ROH]{Na, liq NH_3}$

(g) $CH_3CH{=}CH\overset{O}{\overset{\|}{C}}CH_3 + \underset{1 \text{ mole}}{H_2} \xrightarrow{Pd/BaSO_4}$

(h) 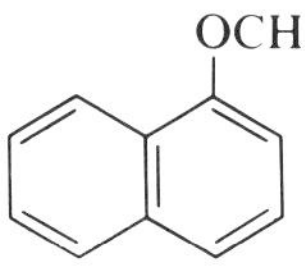 [1-methoxynaphthalene, OCH_3] $\xrightarrow[C_2H_5OH]{Li, NH_3}$

(i) 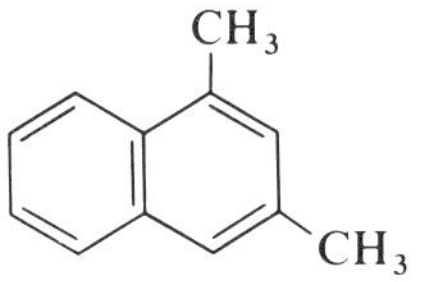 [1,3-dimethylnaphthalene, CH_3, CH_3] $+ 2H_2 \xrightarrow{Pt}$

4. In the preparation of isomycomycin shown on p. 31, what other two organic products would you expect to find from the coupling reaction? What is the intermediate formed between the coupling product shown and isomycomycin?

5. Compound *A* (a monoterpene, $C_{10}H_{16}$) was heated with sulfur at 225° to give Compound *B* ($C_{10}H_{14}$), which when warmed with nitric acid–sulfuric acid gave two isomeric products. Treatment of Compound *B* with sodium in liquid ammonia (trace of ethanol) gave Compound *A* again. Suggest structures for *A* and *B* and also show the three reactions cited.

6. Write balanced equations for the hydrogenation or dehydrogenation reactions in Problems 1(a), 2(d), 2(f), and 3(a).

7. Both chemical reduction and catalytic reduction can involve the use of metals—sodium, zinc, platinum, nickel, etc. Indicate how the two types of reduction differ from one another.

3

Oxygenation of Hydrocarbons

In Chapter 2, we saw that sulfur and selenium are effective in dehydrogenating hydrocarbons, but that oxygen, the most reactive element in this group of the periodic table, usually leads to combustion or gives products that contain oxygen. Although molecular oxygen usually oxidizes a hydrocarbon to a variety of oxygen-containing products with rather little specificity, oxidizing agents containing oxygen can be specific in the products they form from individual classes of organic compounds. In the present chapter, we shall discuss the selective oxidation of hydrocarbons to compounds containing oxygen.

CONVERSION OF ALKENES TO α-GLYCOLS

Of the several classes of hydrocarbons, certainly the easiest to oxidize are the alkenes. Since decolorization of a solution of potassium permanganate is a qualitative test for carbon-carbon double bonds, we can begin our discussion with this reagent. The simplest oxidation of an alkene by permanganate gives an α-glycol, a 1,2-diol. However, the reaction does not usually give good yields, since it is difficult to stop the oxidation at the diol stage. Other products—ketones and hydroxy ketones, aldehydes, acids—are also formed from 1,2-disubstituted alkenes.

The diversity of permanganate oxidation can be illustrated for oleic acid, which in basic solution gives a glycol and in neutral solution a hydroxy ketone.

$$\underset{\text{oleic acid}}{CH_3(CH_2)_7CH{=}CH(CH_2)_7COOH} \xrightarrow[NaOH]{KMnO_4} \underset{erythro}{CH_3(CH_2)_7CH(OH){-}CH(OH)(CH_2)_7COOH}$$

$$\text{oleic acid} \xrightarrow[\text{neutral}]{KMnO_4} CH_3(CH_2)_7\overset{O}{\overset{\|}{C}}{-}\overset{OH}{\overset{|}{C}}H(CH_2)_7COOH + CH_3(CH_2)_7\overset{OH}{\overset{|}{C}}H{-}\overset{O}{\overset{\|}{C}}(CH_2)_7COOH$$

Similarly, in basic solution, bicyclo[2.2.1]heptene (norbornene) gives a glycol with permanganate, whereas in neutral solution a higher oxidation product—in this case a dialdehyde—is formed.

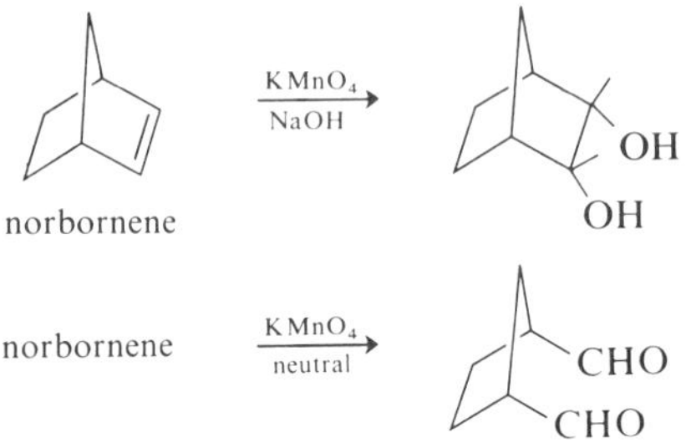

norbornene $\xrightarrow[NaOH]{KMnO_4}$ OH OH

norbornene $\xrightarrow[neutral]{KMnO_4}$ CHO CHO

Although the yields of diols are rarely outstanding with potassium permanganate, the reagent does have the virtue of being cheap, and it is often used. The diol products are nearly always cis diols, as illustrated by both oleic acid and norbornene, and are formed from the less hindered side of the molecule, as illustrated by norbornene. A five-membered cyclic intermediate is postulated to explain the observed stereospecificity. If the reaction is carried out in ^{18}O-labeled water, no label is incorporated into the diol, showing that the hydroxyl groups come from the permanganate rather than the water.

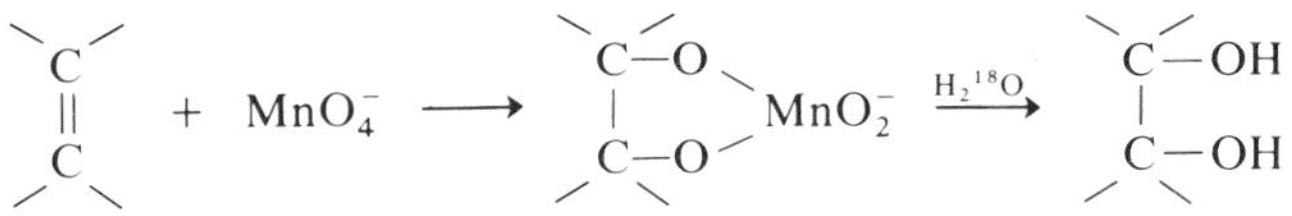

$$\mathrm{C{=}C} + MnO_4^- \longrightarrow \mathrm{C{-}O \atop C{-}O} \rangle MnO_2^- \xrightarrow{H_2{}^{18}O} \mathrm{C{-}OH \atop C{-}OH}$$

Another reagent that reacts in much the same manner as potassium permanganate in forming a cis glycol via a cyclic intermediate is osmium tetroxide. In this case, the cyclic intermediate can be isolated. Hydrolysis with $H_2{}^{18}O$ again shows that the glycol oxygen atoms come from the oxidant, as shown in the equation below. Yields with this reagent are very much better than those with potassium permanganate, and the reaction is more predictable.

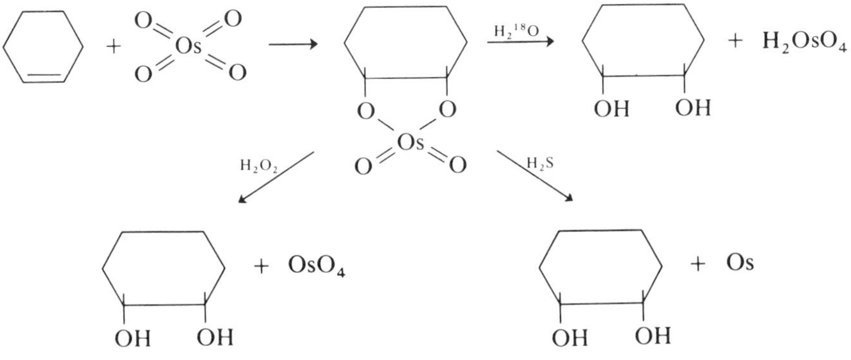

cyclohexene + OsO_4 $\longrightarrow$ cyclic osmate ester (O, O, Os, O, O) $\xrightarrow{H_2{}^{18}O}$ cis-diol (OH OH) + H_2OsO_4

$\xrightarrow{H_2O_2}$ cis-diol (OH OH) + OsO_4

$\xrightarrow{H_2S}$ cis-diol (OH OH) + Os

An osmium tetroxide oxidation can be run in at least three different ways, as illustrated here: the intermediate can be isolated and then hydrolyzed, in which case the inorganic product is an Os^{VI} derivative; or the intermediate can be isolated and reduced with hydrogen sulfide to give metallic osmium; or the alkene can be treated with a catalytic amount of osmium tetroxide and an excess of hydrogen peroxide. The cyclic intermediate formed in the third method is oxidized rapidly by the excess peroxide to release osmium tetroxide, leaving the cis glycol as the organic product. The last method is the procedure of choice, since it has the distinct advantage of limiting the amount of the quite toxic and very expensive osmium tetroxide employed.

Still another method of converting an alkene to a 1,2-diol involves reaction of the carbon-carbon double bond with some reagent to give an intermediate epoxide, which is then opened to give the diol. Reagents utilized for this conversion are usually peracids, which cleave at the oxygen–oxygen bond with transfer of the second oxygen atom to the alkene. Perbenzoic acid is normally employed in organic solvents and gives an epoxide which is isolable. With performic acid in formic acid solvent, the epoxide cannot be isolated, since in the polar solvent the strained three-membered ring is cleaved to give a formate of the 1,2-diol.

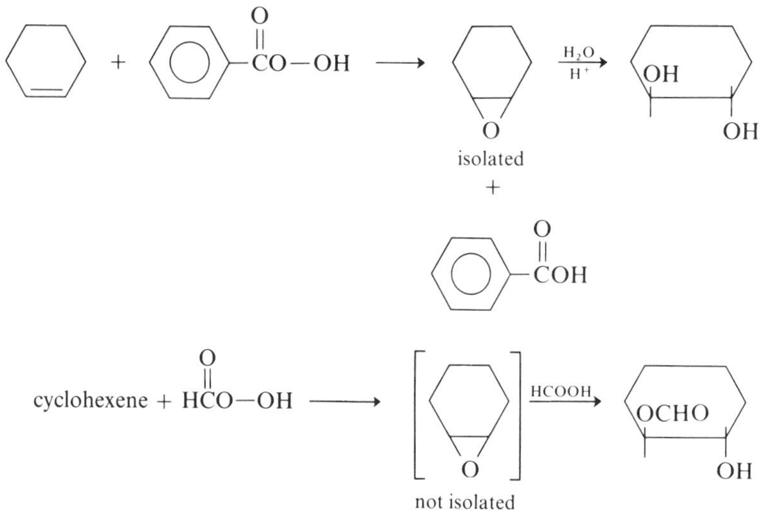

cyclohexene + $C_6H_5CO-OOH$ ⟶ cyclohexene oxide (isolated) + C_6H_5COOH; cyclohexene oxide $\xrightarrow[H^+]{H_2O}$ 1,2-diol (OH, OH)

isolated

cyclohexene + HCO—OH ⟶ [epoxide] (not isolated) $\xrightarrow{HCOOH}$ OCHO / OH

not isolated

The mechanism of epoxidation probably involves a simultaneous transfer of the oxygen and hydrogen of the peracid's hydroxyl group, as shown. The bonds being formed are dotted.

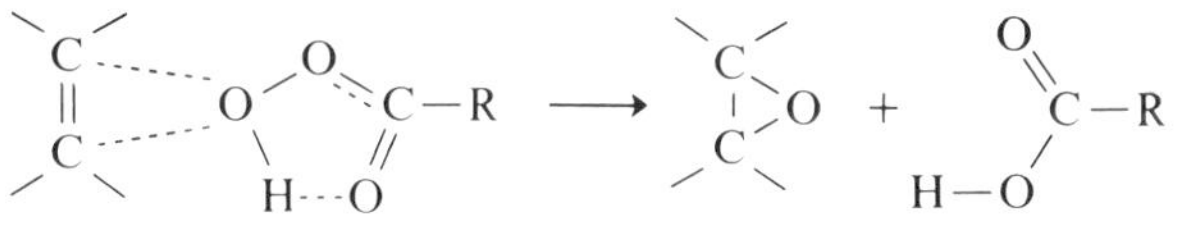

C=C ··· O(—O—C(=O)—R, H···O) ⟶ C—C epoxide (O) + O=C(—R)—O—H

In all these epoxide formations, the oxygen-oxygen bond of the peracid must cleave. The ease of this cleavage is determined by the electron-withdrawing power of the substituent attached to the carbonyl group; thus, pertrifluoroacetic acid is an especially powerful epoxidizing agent owing to the electron-withdrawing fluorine atoms. With this acid, it is possible to form epoxides even from deactivated alkenes, such as conjugated unsaturated esters. To prevent ring opening of the epoxide, the reaction is run in a buffered solution.

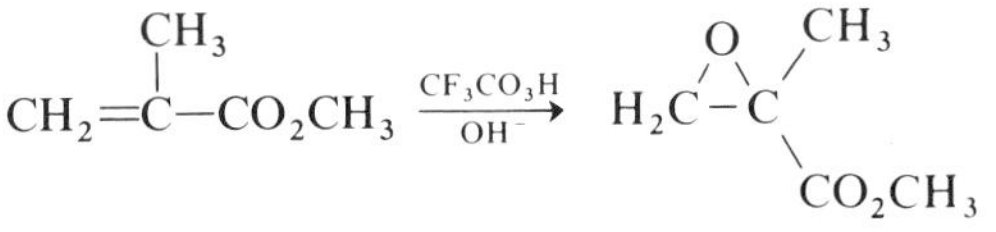

$$CH_2{=}C(CH_3){-}CO_2CH_3 \xrightarrow[OH^-]{CF_3CO_3H} H_2C\overset{O}{-}C(CH_3)(CO_2CH_3)$$

In all reactions of this type, in which an epoxide is first formed and then opened, the glycol isolated is the trans glycol, since the nucleophile that reacts with the epoxide approaches from the side opposite the oxygen atom. In rigid cyclohexene systems, the epoxidizing agent approaches from the less hindered side and the epoxide is preferentially opened to give the trans diaxial product.

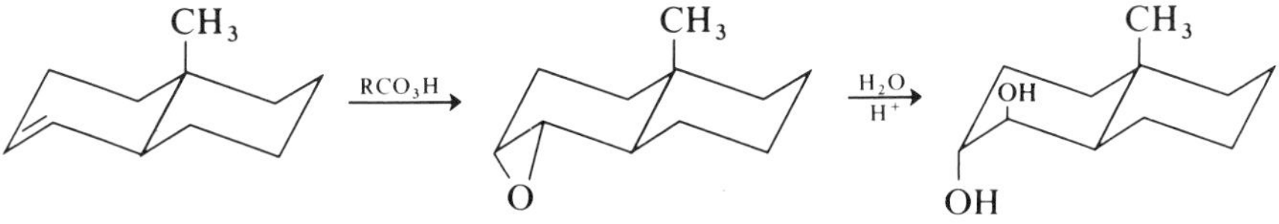

CH_3 $\xrightarrow{RCO_3H}$ CH_3 (O) $\xrightarrow[H^+]{H_2O}$ CH_3 OH OH

Still another method of forming glycols from alkenes involves the use of iodine and silver acetate. The first intermediate in this case is a cyclic iodonium ion, which is formed together with a molecule of silver iodide. Opening of the iodonium ion by acetate gives an acetoxy iodide, and the acetate serves as a neighboring group to displace the adjacent iodide ion, giving a resonance stabilized cyclic acetoxonium ion. This cyclic intermediate then opens, but the direction of opening differs depending upon whether the solvent is dry or contains a trace of water. In the dry solvent, acetate ion serves as a nucleophile, displacing oxonium from the opposite side to give a trans diacetate. In slightly wet solvent, water adds to the acetoxonium ion, giving a cis 1-hydroxy-2-acetate. In the overall reaction, the oxidant is iodine.

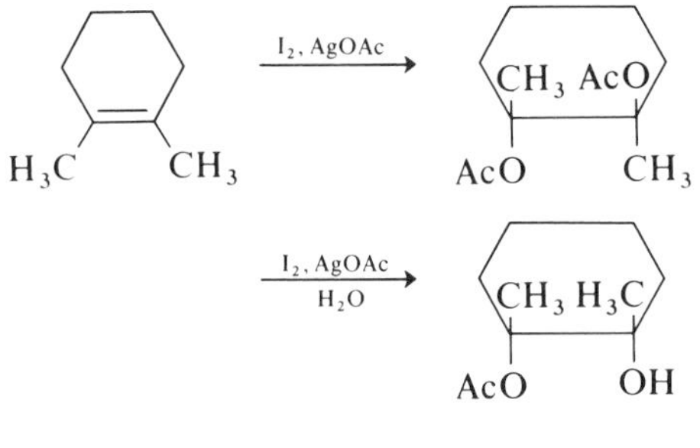

H_3C CH_3 $\xrightarrow{I_2,\ AgOAc}$ CH_3 AcO, AcO CH_3

$\xrightarrow[H_2O]{I_2,\ AgOAc}$ CH_3 H_3C, AcO OH

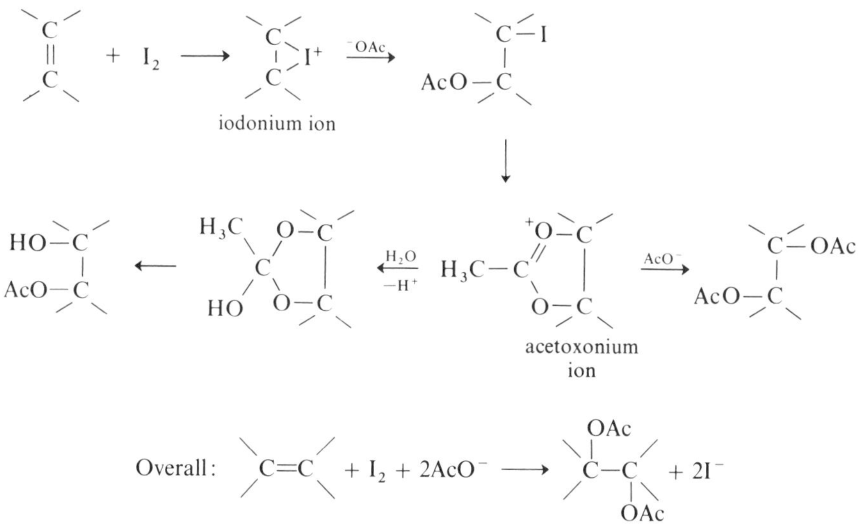

iodonium ion

acetoxonium ion

Overall: $>C{=}C< + I_2 + 2AcO^- \longrightarrow >C(OAc){-}C(OAc)< + 2I^-$

Another reagent that can convert an alkene linkage to a diol, lead tetraacetate, is far less selective than most of the others we have discussed. Since a trans glycol is formed in dry solvent but a cis glycol is formed in slightly wet solvent, an intermediate acetoxonium ion, such as that from iodine-silver acetate, has peen postulated.

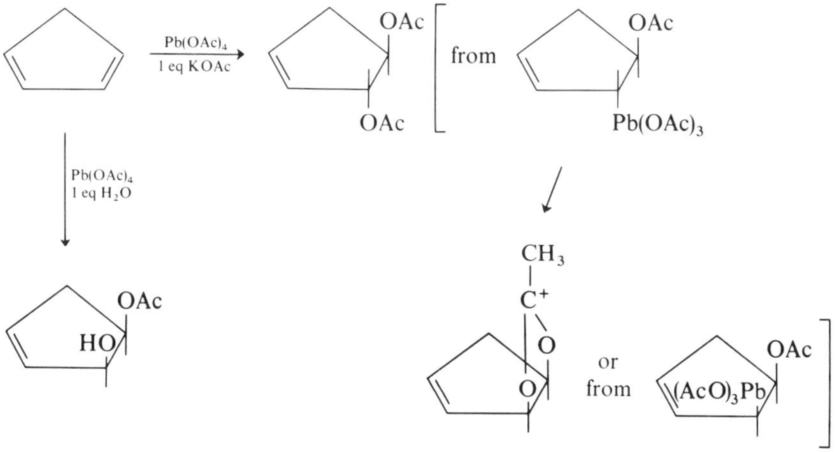

Pb(OAc)$_4$, 1 eq KOAc

from

Pb(OAc)$_4$, 1 eq H_2O

or from

A study carried out to determine the products of oxidation of menthene provides a good illustration of many of the points made in this section. Oxidation with osmium tetroxide gives the cis glycol (*A*) formed when the bulky osmium tetroxide molecule approaches menthene from the side opposite the isopropyl group. Oxidation with iodine and silver acetate in slightly wet acetic acid also gives a cis glycol, actually a mixture of that with both hydroxyl groups cis to methyl (*A*) and that with both cis to isopropyl (*B*). Peracetic acid oxidation gives both trans glycols, *C* and *D*, but mainly the isomer with axial hydroxyl groups (*C*) since the intermediate epoxide opens with water to give preferentially the diaxial trans glycol. With perbenzoic acid, the epoxide can be isolated as a mixture of both isomers, and both open to give nearly exclusively *C*.

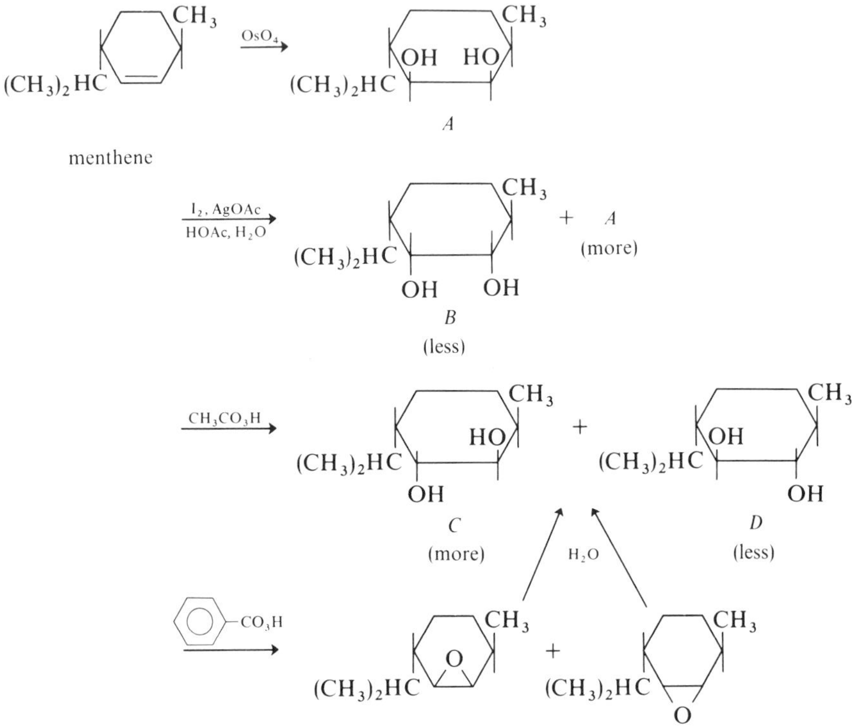

CH_3 $(CH_3)_2HC$ $\xrightarrow{OsO_4}$ CH_3 OH HO $(CH_3)_2HC$

A

menthene

$\xrightarrow[HOAc,\ H_2O]{I_2,\ AgOAc}$ CH_3 $(CH_3)_2HC$ OH OH + *A* (more)

B (less)

$\xrightarrow{CH_3CO_3H}$ CH_3 HO $(CH_3)_2HC$ OH + CH_3 OH $(CH_3)_2HC$ OH

C (more) H_2O *D* (less)

C_6H_5–CO_3H $\longrightarrow$ CH_3 O $(CH_3)_2HC$ + CH_3 $(CH_3)_2HC$ O

Alkenes can be oxidized to α-diketones, as well as to vic-glycols. The reagent used to effect this conversion is selenium dioxide. Unfortunately, selenium dioxide is not very selective and can oxidize other groups, too, as we shall see later in this chapter. In this respect, it resembles lead tetraacetate.

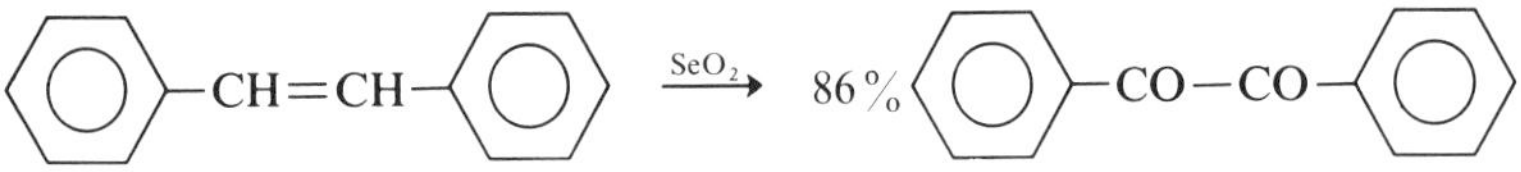

$$C_6H_5\text{–}CH{=}CH\text{–}C_6H_5 \xrightarrow{SeO_2} 86\%\ C_6H_5\text{–}CO\text{–}CO\text{–}C_6H_5$$

CLEAVAGE OF ALKENES BY OXIDATION

We noted in the preceding section that it is relatively difficult to stop the oxidation of alkenes by potassium permanganate at the diol stage, although we can sometimes do so by working under controlled conditions in basic solution. We also saw that oxidation of norbornene with permanganate in neutral solution gave a dialdehyde, thus effecting a cleavage of its double bond. It is unusual for permanganate oxidation to give aldehydes, the more common products from cleavage of 1,2-disubstituted alkenes being acids. For example, ricinoleic acid, the principal constituent of castor oil, can be oxidized by potassium permanganate to azelaic acid in approximately 35% yield.

$$CH_3(CH_2)_5\overset{OH}{\underset{|}{C}}HCH_2\overset{H}{\underset{|}{C}}{=}\overset{H}{\underset{|}{C}}(CH_2)_7COOH \xrightarrow[\text{dil KOH, }75^\circ]{KMnO_4} 35\%\ HOOC(CH_2)_7COOH$$

Depending upon the degree of substitution at the ends of the double bond, the product may be an acid, a ketone, or carbon dioxide, as is illustrated in the oxidation of the cyclohexene derivative shown. However, the yields in the permanganate reaction are not good, largely because permanganate is a relatively nonspecific reagent, and further oxidation occurs, resulting in acids of shorter chain length.

$$\text{1-cyclohexenyl-}(CH_2)_4CH{=}CH_2 \xrightarrow[\text{acetone, r.t.}]{KMnO_4} 11\%\ HOOC(CH_2)_4CO(CH_2)_4COOH$$

$$\text{cyclohexene} \xrightarrow{KMnO_4} HOOC(CH_2)_4COOH + HOOC(CH_2)_nCOOH \quad n \leq 3$$

Other strong oxidizing agents can also be used to cleave the carbon-carbon double bond. Chromic oxide in acidic solution is frequently employed. For this purpose, acetic acid is somewhat better than sulfuric acid, since it gives less rearrangement to side products. A simple example is the oxidation of tetraphenylethylene to benzophenone, as shown. Apparently, an epoxide is an intermediate in this reaction because it may be isolated if only a limited amount of chromic oxide is employed.

$$(C_6H_5)_2C{=}C(C_6H_5)_2 \xrightarrow[HOAc]{CrO_3} \text{epoxide } (C_6H_5)_2\overset{O}{C{-}C}(C_6H_5)_2 \xrightarrow[HOAc]{\text{excess } CrO_3} C_6H_5{-}CO{-}C_6H_5$$

A more complicated example is the oxidation of tetramethylethylene with chromic oxide and sulfuric acid, which leads to a mixture of products. In addition to the expected acetone, pinacolone and pivalic acid are both obtained, as well as the epoxide of tetramethylethylene. A suggested intermediate is the carbonium ion shown, since it could rearrange to an epoxide or to pinacolone or give the glycol, which on oxidation would yield acetone. Pivalic acid could be obtained from further oxidation of pinacolone.

$$(CH_3)_2C{=}C(CH_3)_2 \longrightarrow \underset{}{CH_3COCH_3} + \underset{\text{pinacolone}}{(CH_3)_3COCH_3} + \underset{\text{pivalic acid}}{(CH_3)_3COOH} + (H_3C)_2C\overset{O}{-}C(CH_3)_2$$

$$(CH_3)_2C{=}C(CH_3)_2 \xrightarrow{HOCrO_2^+} \left[(CH_3)_2\overset{+}{C}{-}C(CH_3)_2{-}O{-}Cr(=O)OH \right] \longrightarrow \text{epoxide};\ \longrightarrow \text{pinacolone} \longrightarrow \text{pivalic acid}$$

$$\downarrow$$

$$\left[CH_3{-}\underset{OH}{\overset{CH_3}{C}}{-}\underset{OH}{\overset{CH_3}{C}}{-}CH_3 \right] \longrightarrow \text{acetone}$$

Clearly, from the difficulties found in the oxidation of alkenes by strong oxidizing agents such as potassium permanganate and chromic oxide, a more specific reagent is needed for cleavage of the alkene linkage. Fortunately, two specific and useful reagents are available for this purpose. One is a combination of sodium periodate and potassium permanganate (or sodium periodate and osmium tetroxide), which we shall discuss in Chapter 4, since it is more clearly related to periodate oxidations. The other is a reagent that has found enormous use in structure determination of alkenes, as well as in analysis and synthesis. This reagent is ozone, O_3. Ozone is generated by passing an electric discharge through oxygen, and machines for this purpose are available in most organic laboratories. Recently, ozone has also become available as a gas in cylinders, dissolved in solutions of Freon.

The reaction of ozone with alkenes can be illustrated with 3-hexene. If ozonized oxygen is passed through a solution of 3-hexene, the ozone is absorbed very rapidly to give a quantitative yield of a product, which is called an ozonide, with the composition $C_6H_{12}O_3$. The ozonide can be isolated and upon hydrolysis gives a mixture of propionic acid and propionaldehyde.

$$CH_3CH_2CH{=}CHCH_2CH_3 + O_3 \longrightarrow \underset{\text{ozonide}}{C_6H_{12}O_3} \xrightarrow{H_2O} CH_3CH_2CHO + CH_3CH_2COOH$$

The ozonolysis shown is not completely satisfactory since a mixture of products is obtained. However, the problem has two solutions: one is to oxidize the initial ozonide, and the other is to reduce it. In the oxidative work-up of an ozonolysis reaction, hydrogen peroxide is generally employed; in the ozonolysis of 3-hexene, this gives 2 moles of propionic acid. The alternative, reductive, work-up can be carried out with zinc or other chemical reducing agents, or, more conveniently, with hydrogen over a metal catalyst at atmospheric pressure. The product of reduction of the ozonide of 3-hexene is propionaldehyde.

$$\text{Oxidative work-up} \qquad C_6H_{12}O_3 \xrightarrow{H_2O_2} 2CH_3CH_2COOH$$

$$\text{Reductive work-up} \qquad C_6H_{12}O_3 \xrightarrow[Pd]{H_2} 2CH_3CH_2CHO$$

If the alkene undergoing ozonolysis is disubstituted on one end, a ketone is obtained, as illustrated by the ozonolysis of the isomeric hexene, tetramethylethylene.

$$(CH_3)_2C{=}C(CH_3)_2 \xrightarrow{O_3} \underset{\text{ozonide}}{C_6H_{12}O_3} \xrightarrow[Pd]{H_2} 2CH_3COCH_3$$

One of the principal uses of ozonolysis is as an aid in deducing the structures of alkenes, since oxygen is introduced on all the carbon atoms that in the starting material contained a double bond. For this purpose, it is generally better to use a reductive work-up of the ozonolysis reaction, because aldehydes and ketones are very easily identified as derivatives such as their dinitrophenylhydrazones. The use of ozonolysis in the structure proof of an alkene can be illustrated in the following example.

A hydrocarbon (A, C_6H_{12}) was treated with ozone, then with hydrogen and palladium, to give a mixture of acetaldehyde and methyl ethyl ketone. By reconstructing the hydrocarbon with the carbon-carbon double bond in the position occupied by the carbon-oxygen double bonds in the products, the structure of A can readily be shown to be 3-methyl-2-pentene. Of course, cis and trans isomers are possible for this structure, but ozonolysis does not provide evidence relating to the stereochemistry.

$$\underset{A}{C_6H_{12}} \xrightarrow{O_3} \xrightarrow[Pd]{H_2} CH_3\overset{\overset{O}{\|}}{C}H + CH_3\overset{\overset{O}{\|}}{C}C_2H_5$$

$$\text{so } A \text{ is } CH_3CH{=}\overset{\overset{CH_3}{|}}{C}{-}C_2H_5.$$

The mechanism of ozonolysis has been a subject of speculation and research for many years. In the preceding discussion, the ozonide of 3-hexene was indicated only by its molecular formula, $C_6H_{12}O_3$. Actually, two ozonides are apparently formed: a very unstable compound which forms first is called a molozonide, and this rearranges to a more stable isomer called an isozonide. In a few cases molozonides have been isolated. That from *trans*-1,2-di-*t*-butylethylene is stable at $-70°$, but rearranges at higher temperatures. The carbon atoms of the alkene must still be bonded to one another, since reduction of the molozonide gives the glycol shown. From studies of the nmr spectrum of this molozonide, the structure was concluded to be that shown, with a five-membered ring containing the terminal oxygen atoms of ozone attached to the two alkene carbon atoms, rather than the alternative containing a four-membered ring. In the rearrangement of the molozonide to the isozonide, an intermediate carbonyl compound and zwitterion are apparently formed. These may either recombine with one another immediately to give the isozonide or diffuse apart to react with other species, depending upon the conditions of the reaction. In alcohol solvents, an alkoxy hydroperoxide is usually formed.

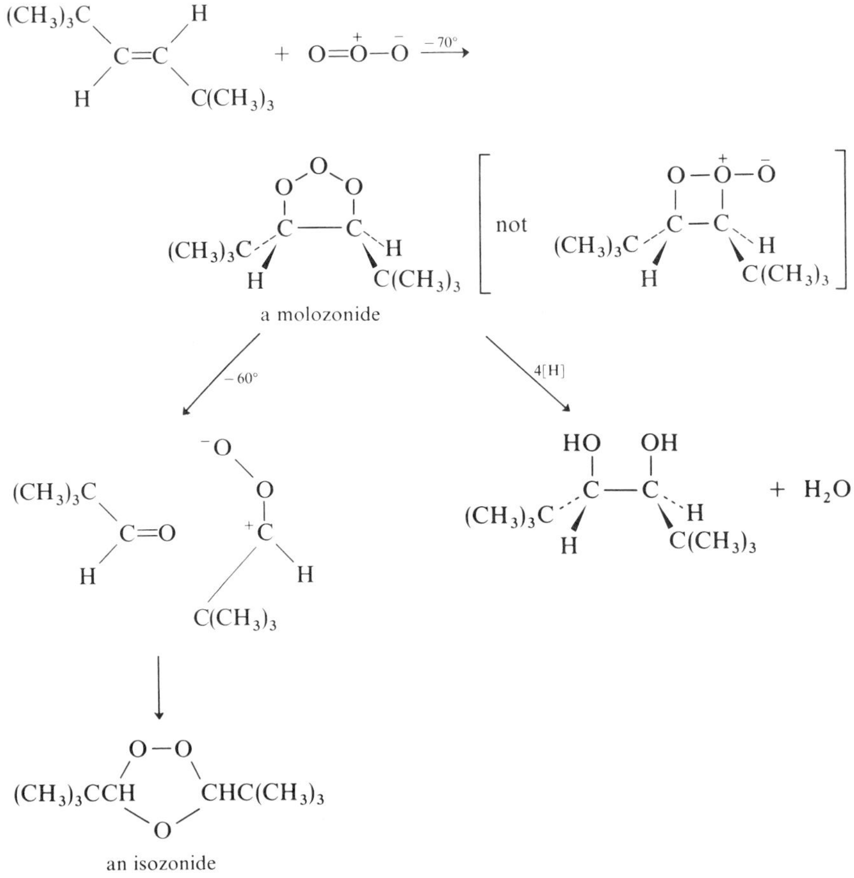

$(CH_3)_3C(H)C{=}C(H)C(CH_3)_3 + O{=}\overset{+}{O}{-}\overset{-}{O} \xrightarrow{-70°}$

a molozonide [not the four-membered ring structure with $O{-}\overset{+}{O}{-}\overset{-}{O}$]

$\xrightarrow{-60°}$ $(CH_3)_3CCH{=}O$ + $^{-}O{-}O{-}\overset{+}{C}H{-}C(CH_3)_3$ $\longrightarrow$ an isozonide, $(CH_3)_3CCH(O{-}O)(O)CHC(CH_3)_3$

$\xrightarrow{4[H]}$ $(CH_3)_3C(H)C(OH){-}C(OH)(H)C(CH_3)_3 + H_2O$

$$(CH_3)_2C{=}C(CH_3)_2 \xrightarrow[ROH]{O_3} (CH_3)_2C{=}O + RO{-}C(OOH)(CH_3){-}CH_3$$

In the absence of hydrophilic solvents (alcohols and acids), the carbonyl component and the zwitterion recombine, but not necessarily with the same starting partners. Thus, the ozonolysis of oleic acid has been shown to give a mixture of six different isozonides (3 pairs of cis-trans isomers).

$$CH_3(CH_2)_7CH{=}CH(CH_2)_7COOH \xrightarrow{O_3} \text{RHC}\overset{O}{\frown}\text{CHR}'(O{-}O) + \text{R}'\text{HC}\overset{O}{\frown}\text{CHR}'(O{-}O) + \text{RHC}\overset{O}{\frown}\text{CHR}(O{-}O) + \text{trans isomers}$$

OXYGENATION OF ALKYNES

The carbon-carbon triple bond reacts in much the same way as the carbon-carbon double bond toward oxidizing agents, just as it does toward electrophilic reagents in general. However, oxidations of alkynes are usually more complex than those of alkenes and are seldom as useful. Although cleavage to acid products is the general route, it is sometimes possible to stop at an intermediate stage. An example involves the oxidation of the acetylenic analogue of oleic acid to the corresponding di-keto acid by treatment with potassium permanganate under essentially neutral conditions.

$$CH_3(CH_2)_7C{\equiv}C(CH_2)_7COOH \xrightarrow[pH\ 7.5]{aq\ KMnO_4} CH_3(CH_2)_7\overset{O}{\overset{\|}{C}}{-}\overset{O}{\overset{\|}{C}}(CH_2)_7COOH$$

Like many other reactions with electrophilic reagents, oxidation of alkynes usually takes place much more slowly than that of alkenes, as illustrated in the following example:

$$HC{\equiv}C(CH_2)_7CH{=}C(C_6H_5)_2 \xrightarrow[HOAc]{CrO_3} HC{\equiv}C(CH_2)_7COOH + (C_6H_5)_2CO$$

OXIDATION OF SATURATED HYDROCARBONS

We think of saturated hydrocarbons as being completely inert or saturated to most reagents, as indeed they are. It is this lack of reactivity that makes saturated hydrocarbons excellent as solvents in many reactions. When they do react, it is usually under extreme conditions, such as radical chlorination or combustion to provide energy for automobile and diesel engines. This can be illustrated by the burning of "isooctane" to give carbon dioxide and water.

$$CH_3C(CH_3)_2CH_2CH(CH_3)CH_3 + 12\tfrac{1}{2}O_2 \xrightarrow{\Delta\Delta} 8CO_2 + 9H_2O$$

However, there is some selectivity in oxidation of saturated hydrocarbons, even by air, and much of the huge petrochemical industry is based on the controlled oxidation of saturated hydrocarbons to give such useful products as methanol, formaldehyde, and other oxygenated materials. These, however, are industrial processes requiring specific catalysts and conditions and will not be discussed here. They are treated in the book in this series by Stille.*

Some preferential reactivity is also shown in the reactions of saturated hydrocarbons with laboratory reagents, such as permanganate and, especially, chromic oxide. Tertiary hydrogen atoms are attacked more rapidly than secondary hydrogens, and these, in turn, are attacked more rapidly than primary hydrogens. The products, however, are not alcohols, although alcohols may be intermediates that are oxidized further. Products isolated include ketones and acids, and selective oxidation in chromic oxide-acetic acid has been used to locate branching in saturated hydrocarbons by identification of these ketones and acids, as illustrated for the oxidation of 10-methyloctadecanoic acid.

$$CH_3(CH_2)_7CH(CH_3)(CH_2)_8COOH \xrightarrow[HOAc \; 65\%]{CrO_3} 1\%\ CH_3(CH_2)_7COCH_3 + 10\%\ CH_3(CH_2)_6COOH + 22\%\ HOOC(CH_2)_7COOH$$

Another illustration of preferential oxidation at tertiary and secondary hydrogens compared with primary hydrogens is the Kuhn-Roth determination, a standard semiquantitative organic procedure. This analysis indicates the number of C-methyl groups in a molecule by oxidation with chromic oxide in sulfuric acid and distillation of the product. For every mole of

* J. K. Stille, *Industrial Organic Chemistry*, in Foundations of Modern Organic Chemistry Series, Prentice-Hall, Inc., Englewood Cliffs, N. J., 1968.

acetic acid formed, there must have been at least 1 mole of C-methyl present; *n*-octane, for example, could give a maximum of 2 moles of acetic acid.

$$CH_3(CH_2)_6CH_3 \xrightarrow[H_2SO_4]{CrO_3} \xrightarrow{\text{distill}} \leq 2CH_3COOH$$

This maximum value is never reached, but anything above 1 mole is indicative of more than one C-methyl group. Kuhn-Roth oxidation of the 10-methyloctadecanoic acid of the preceding paragraph would also give a maximum of 2 moles of acetic acid. The number of moles of acid in the distillate is estimated by titration, and the volatile acids can be identified more specifically by paper chromatography. Identification of propionic acid, for example, would indicate a C-ethyl group (as well as a C-methyl group).

ALLYLIC AND BENZYLIC OXIDATION

We have just seen that saturated hydrocarbons are rather unreactive toward oxidizing agents. Although there is a general preference in the order tertiary > secondary > primary, this reactivity difference is often not great. However, a carbon-hydrogen bond adjacent to an alkene linkage or a benzene ring is especially susceptible to oxidation, and many reagents react in a highly selective manner at these positions. Most of the oxidations are presumed to involve radical intermediates—alkyl or benzyl radicals that are highly stabilized by resonance, although the corresponding allyl or benzyl cations from two-electron oxidations would also be resonance stabilized. A very important example of selective oxidation in the benzylic position is the industrially useful chain reaction of oxygen with cumene (isopropylbenzene), which proceeds by the mechanism shown and gives cumene hydroperoxide in excellent yield. The mechanism involves the usual initiation, propagation, and termination steps of chain reactions, as shown.

$$C_6H_5-CH(CH_3)_2 + O_2 \longrightarrow C_6H_5-C(CH_3)_2-OOH$$

Initiation: $RH + O_2 \longrightarrow R\cdot + HOO\cdot$

Propagation: $R\cdot + O_2 \longrightarrow R-O-O\cdot$

$R-O-O\cdot + RH \longrightarrow R-O-OH + R\cdot$

Termination: $R\cdot + R-O-O\cdot \longrightarrow R-O-O-R$

$R\cdot + HOO\cdot \longrightarrow R-O-OH$

$R\cdot + R\cdot \longrightarrow R-R$

Tetralin reacts similarly to give tetralin hydroperoxide. Radical reactions of oxygen are discussed in considerable detail in the book in this series by Pryor.*

$$\text{Tetralin} \xrightarrow{O_2} \text{Tetralin hydroperoxide (OOH)}$$

Although reactions with oxygen are mainly of interest in industry, a number of laboratory reagents are also highly selective in attacking an alpha position, activated by either an adjacent carbon-carbon double bond, a benzene ring, or a carbonyl group. We have already encountered some of these reagents in Chapter 2, in which we noted that carbon-carbon double bonds can be introduced adjacent to existing double bonds by selenium dioxide and manganese dioxide. The more common mode of reaction of selenium dioxide, however, is the introduction of a carbonyl group adjacent to unsaturation. Thus, selenium dioxide oxidizes cyclohexanone to cyclohexanedione and diphenylmethane to benzophenone.

$$\text{Cyclohexanone} + SeO_2 \xrightarrow[\text{water}]{\text{dioxane}} 55\%\ \text{1,2-cyclohexanedione} + Se + H_2O$$

$$C_6H_5{-}CH_2{-}C_6H_5 \xrightarrow{SeO_2} 87\%\ C_6H_5{-}CO{-}C_6H_5$$

Selenium dioxide can also introduce hydroxyl groups: oxidation of 1-heptyne gives 1-heptyn-3-ol, whereas oxidation of cyclohexene gives a mixture of cyclohexenone and cyclohexenol.

$$CH_3(CH_2)_4C{\equiv}CH \xrightarrow{SeO_2} 27\%\ C_4H_9CH(OH)C{\equiv}CH$$

$$\text{Cyclohexene} \xrightarrow{SeO_2} 50\%\ \text{cyclohexenol (OH)} + 6\%\ \text{cyclohexenone (O)}$$

The mechanism of selenium dioxide oxidation has been studied in detail for the oxidation of deoxybenzoin to benzil. The rate-determining step apparently is the formation of a selenite ester of the enol form, which then

* W. A. Pryor, *Introduction to Free Radical Chemistry*, in Foundations of Modern Organic Chemistry Series, Prentice-Hall, Inc., Englewood Cliffs, N. J., 1966.

decomposes rapidly to give, successively, the selenium(II) ester of benzoin, then benzil plus elemental selenium.

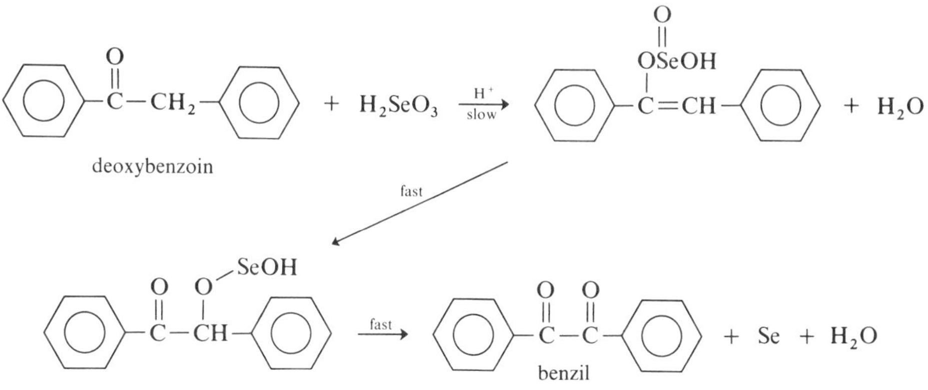

deoxybenzoin + H_2SeO_3 $\xrightarrow[\text{slow}]{H^+}$ + H_2O

fast

fast

+ Se + H_2O

benzil

In contrast to selenium dioxide, which most often gives carbonyl products, mercuric acetate and lead tetraacetate both usually introduce acetoxyl groups in the alpha position. Any position activated by a carbon-carbon or carbon-oxygen double bond is susceptible to attack.

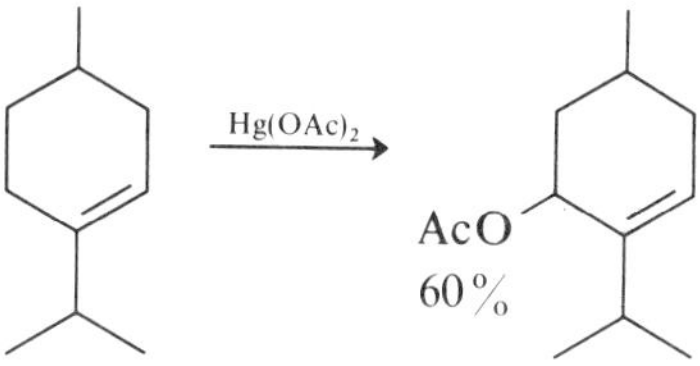

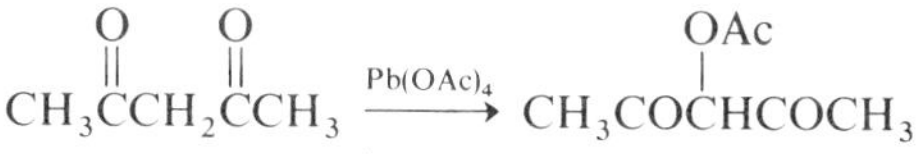

$\xrightarrow{Hg(OAc)_2}$ AcO 60%

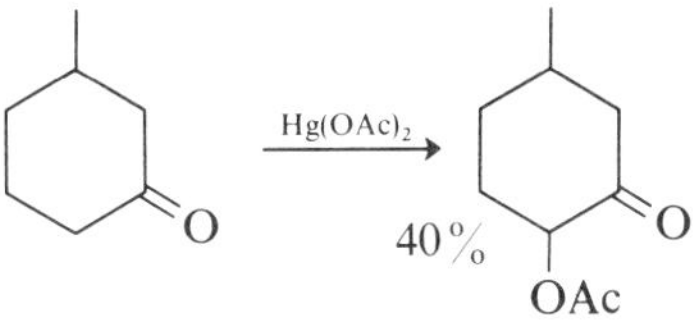

$\xrightarrow{Hg(OAc)_2}$ 40% OAc

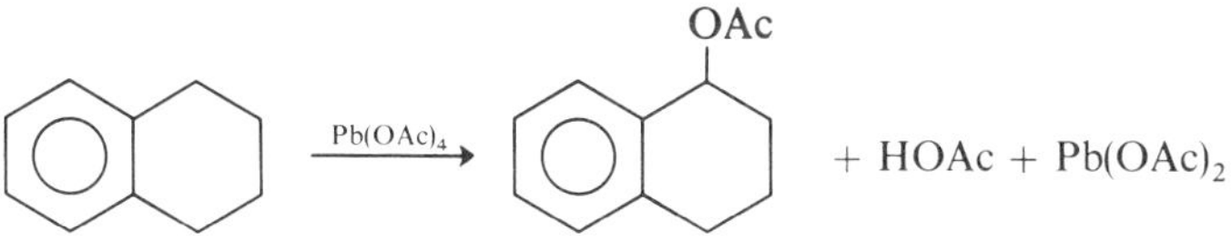

$\xrightarrow{Pb(OAc)_4}$ OAc + HOAc + $Pb(OAc)_2$

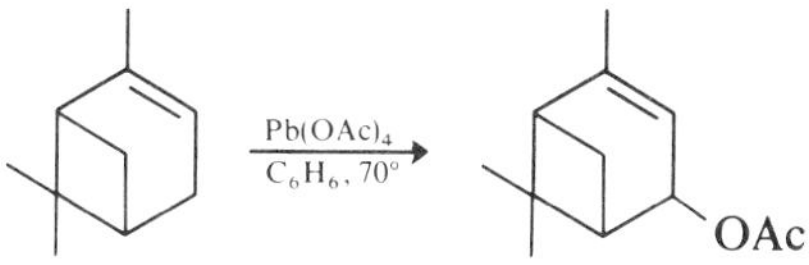

$\xrightarrow[C_6H_6,\ 70^\circ]{Pb(OAc)_4}$ OAc

$$CH_3\overset{O}{\overset{\|}{C}}CH_2\overset{O}{\overset{\|}{C}}CH_3 \xrightarrow{Pb(OAc)_4} CH_3CO\overset{OAc}{\overset{|}{C}H}COCH_3$$

OXIDATION OF AROMATIC HYDROCARBONS

Like saturated aliphatic hydrocarbons, aromatic hydrocarbons are usually regarded as being good solvents because they are chemically inert to oxidizing reagents. On a comparative basis, this is perhaps true in that functional groups on aromatic compounds can usually be selectively oxidized. For example, styrene on ozonolysis gives, in excellent yield, benzaldehyde and formaldehyde. When treated with potassium permanganate, the same compound would give benzoic acid and carbon dioxide.

$$C_6H_5{-}CH{=}CH_2 \xrightarrow{O_3} \xrightarrow[Pd]{H_2} C_6H_5{-}CHO + CH_2O$$

On the other hand, special conditions are available, involving vanadium pentoxide as an oxidant, which convert benzene to maleic anhydride in sufficiently good yield to make this a commercial process, and the same high temperature process oxidizes naphthalene to phthalic anhydride.

$$C_6H_6 + \tfrac{9}{2}O_2 \xrightarrow[425^\circ]{V_2O_5} \text{maleic anhydride} + 2H_2O + 2CO_2$$

$$C_{10}H_8 + \tfrac{9}{2}O_2 \xrightarrow[400^\circ]{V_2O_5} \text{phthalic anhydride} + 2H_2O + 2CO_2$$

Milder conditions can usually be used to introduce oxygen into one of the aromatic rings of condensed ring aromatic compounds, since one ring is often very susceptible to oxidation. Naphthalene, for example, gives naphthoquinone (together with phthalic anhydride and other products) with chromic oxide, anthracene gives anthraquinone, and phenanthrene gives phenanthraquinone.

$$\text{naphthalene} \xrightarrow{CrO_3} \text{1,4-naphthoquinone} + \text{other products}$$

$$\text{anthracene} \xrightarrow{CrO_3} 91\%\ \text{anthraquinone}$$

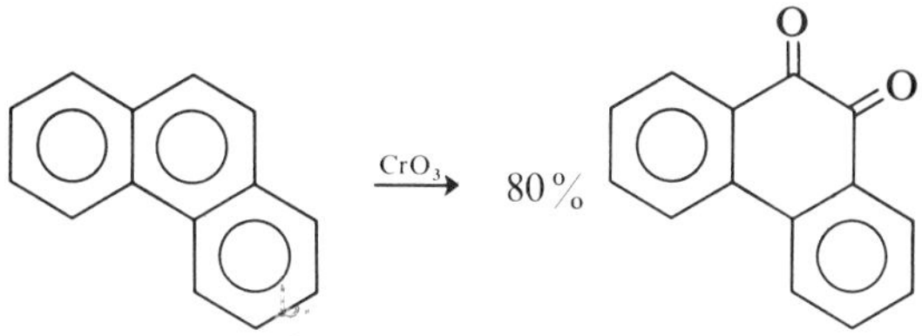

$$\xrightarrow{CrO_3} \quad 80\%$$

Ozone reacts preferentially with carbon-carbon double bonds, but it also reacts with aromatic rings, and this reagent was, in fact, used to demonstrate that two equivalent canonical formulas (resonance forms) are required to describe a benzene ring. Ozonolysis of *o*-xylene, for example, gives approximately 3 moles of glyoxal (2 moles from one canonical formula and 1 from the other), 2 moles of pyruvaldehyde, and 1 mole of biacetyl.

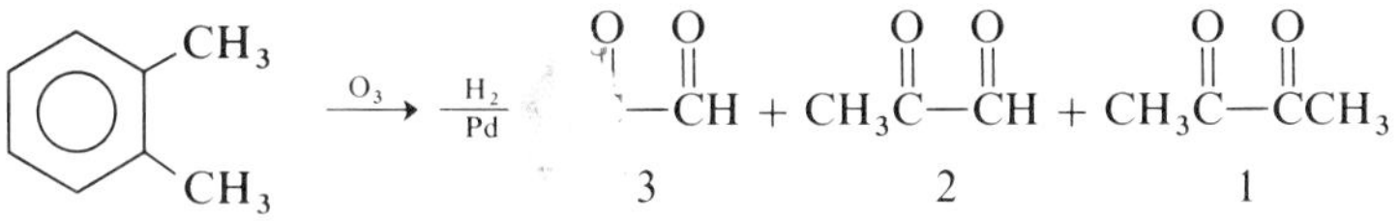

$$C_6H_4(CH_3)_2 \xrightarrow{O_3} \xrightarrow[Pd]{H_2} \underset{3}{HC(=O)-CH(=O)} + \underset{2}{CH_3C(=O)-CH(=O)} + \underset{1}{CH_3C(=O)-C(=O)CH_3}$$

With condensed ring aromatic hydrocarbons, ozone can be quite selective in reacting preferentially at the most electron-rich double bonds, since it is an electrophilic reagent. Attempts have been made to correlate the selectivity of ozonolysis with other properties of condensed ring hydrocarbons, such as the position of electrophilic substitution, and even with the degree of carcinogenicity (ability to induce cancer).

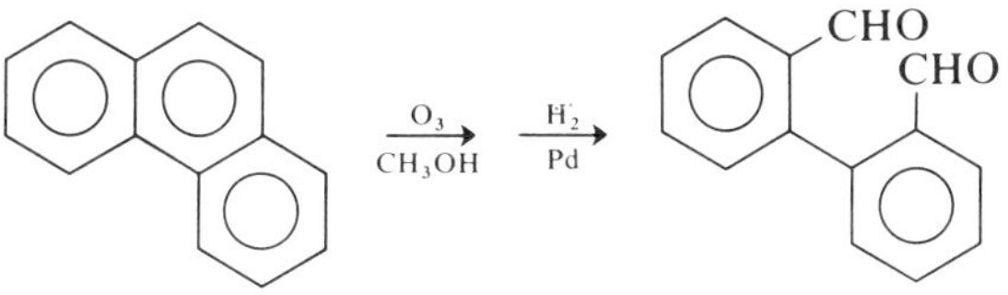

$$\xrightarrow[CH_3OH]{O_3} \xrightarrow[Pd]{H_2}$$

CHO CHO

OXIDATION OF ALKYL AROMATIC HYDROCARBONS

When two unreactive hydrocarbon portions of the same molecule are present, one of them a saturated aliphatic group and the other an aromatic group, it is very interesting to observe which group reacts toward specific oxidizing agents. In general, it is the alkyl group that reacts in large measure, because it usually has a benzylic position susceptible to attack by oxidizing agents of both the radical and cationic variety. The only exception is in those compounds such as *t*-butylbenzene, which do not contain hydrogen atoms in the alpha position; they are stable toward most oxidizing agents. Strong oxidizing agents are very effective in attacking the alkyl groups of alkyl aromatic compounds. In fact, oxidation of this sort constitutes a standard method for identifying positional isomers of multisubstituted alkyl benzenes. For example, *o*-chlorotoluene, a liquid, gives *o*-chlorobenzoic acid, a solid, and mesitylene gives mesitoic acid (1,3,5-benzenetricarboxylic acid).

$$o\text{-}ClC_6H_4CH_3 \xrightarrow{KMnO_4} o\text{-}ClC_6H_4COOH$$

$$\underset{\text{mesitylene}}{1,3,5\text{-}(CH_3)_3C_6H_3} \xrightarrow{KMnO_4} \underset{\text{mesitoic acid}}{1,3,5\text{-}(HOOC)_3C_6H_3}$$

Oxidation of the side chain can also be effected with other oxidizing agents, such as chromic oxide and nitric acid; for example, *p*-nitrotoluene with dichromate gives *p*-nitrobenzoic acid in excellent yield. Usually, the conditions employed can be vigorous, since the aromatic ring is stable to oxidation. When substituents relatively sensitive to oxidation are present, however, it is necessary to employ more selective conditions, as in the oxidation of *o*-methylacetanilide to the corresponding carboxylic acid.

$$p\text{-}O_2NC_6H_4CH_3 + Na_2Cr_2O_7 \xrightarrow{H_2SO_4} 86\%\ p\text{-}O_2NC_6H_4COOH$$

$$o\text{-}CH_3C_6H_4NHCOCH_3 \xrightarrow[MgSO_4]{KMnO_4} o\text{-}HOOCC_6H_4NHCOCH_3$$

Something of an exception to the rule of preferential oxidation of alkyl groups is found in the Kuhn-Roth oxidation: some acetic acid is formed from oxidation of *m*-xylene but the yield is low.

$$m\text{-}CH_3C_6H_4CH_3 \xrightarrow[H_2SO_4]{CrO_3} \underset{0.24\text{ mole }(12\%)}{CH_3COOH}$$

Some heterocyclic compounds such as pyridine behave like benzene in being quite stable to oxidation, and nicotinic acid can be obtained by the oxidation of 3-methylpyridine with potassium permanganate.

$$\text{3-methylpyridine} \xrightarrow[H_2O,\ 100°]{KMnO_4} 50\%\ \text{3-pyridinecarboxylic acid (nicotinic acid)} + MnO_2$$

Carboxylic acids are almost the universal products in these oxidations with strong oxidants, since the intermediate alcohols, aldehydes, and ketones are even less stable to oxidation than the starting materials. However, in the Étard reaction, which employs chromyl chloride in place of chromic oxide, the aldehyde is the usual product obtained.

$$\text{2-nitro-4-methoxytoluene } (CH_3,\ NO_2,\ OCH_3) \xrightarrow[CS_2,\ 25°]{CrO_2Cl_2} \xrightarrow{H_2O} 50\%\ \text{2-nitro-4-methoxybenzaldehyde } (CHO,\ NO_2,\ OCH_3)$$

One of the few general methods for oxidizing an aromatic ring in preference to an aliphatic side chain is ozonolysis. Ozone reacts only with π-electron systems, and, thus, gives good yields of products retaining the alkyl group, with a carboxyl group replacing the aromatic ring.

$$C_6H_5CH_2CH_3 \xrightarrow{O_3} \xrightarrow{H_2O_2} CH_3CH_2COOH$$

REFERENCES

Augustine (*Oxidation*): Chap. 6, J. S. Belew, "Ozonization."

Augustine (*Oxidation*): Chap. 1, D. G. Lee, "Hydrocarbon Oxidation Using Transition Metal Compounds."

Augustine (*Oxidation*): Chap. 5, S. N. Lewis, "Peracid and Peroxide Oxidations."

Augustine (*Oxidation*): Chap. 3, E. N. Trachtenberg, "Selenium Dioxide Oxidation."

Bailey, P. S., "Reactions of Ozone with Organic Compounds," *Chem. Rev.*, **58**, 925 (1958).

Bentley (Part 2): Chap. 17, K. W. Bentley, "Carbon-Carbon Double Bond Fission."

Bentley (Part 2): Chap. 16, K. T. Potts, "Carbon-Carbon Single Bond Fission."

Cullis, C. F., and A. Fish, "Carbonyl-Forming Oxidations," in "Chemistry of the Carbonyl Groups," S. Patai, ed., John Wiley & Sons, Inc., New York, 1966, Chap. 2.

Hock, H., and H. Kropf, "Autoxidation of Hydrocarbons," *Angew. Chem.*, **69**, 313 (1957).

House: Chap. 6, "Oxidations with Peracids and Other Peroxides."

Stewart: Chap. 3, "The Breaking of Carbon-Hydrogen and Carbon-Carbon Bonds"; also Chaps. 4, 5, 6, 8, 9.

Waters: Chap. 2, "The Direct Oxidation of C-H Bonds; also Chaps. 3 and 8.

Wiberg: Chap. 5, R. Criegee, "Oxidations with Lead Tetraacetate."

Wiberg: Chap. 1, R. Stewart, "Oxidation by Permanganate."

Wiberg: Chap. 3, W. A. Waters and J. S. Littler, "Oxidation by Vanadium (V), Cobalt (III), and Manganese (III)."

Wiberg: Chap. 2, K. B. Wiberg, "Oxidation by Chromic Acid and Chromyl Compounds."

PROBLEMS

1. Write a detailed mechanism for each of the following reactions, including all intermediates and/or transition states.

(a) 1,2-dimethylcyclohexene (CH_3, CH_3) $\xrightarrow[H_2O]{I_2,\ AgOAc}$ cyclohexane with CH_3, H_3C, AcO, OH

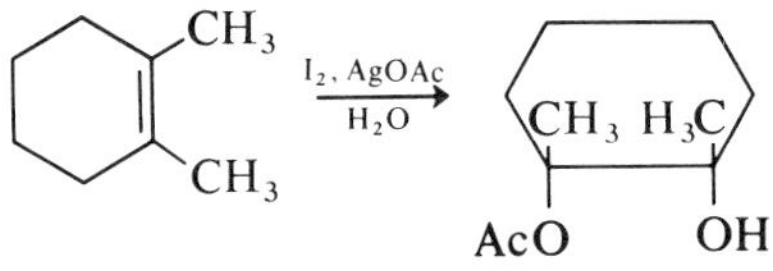

(b) $(CH_3)_2C{=}C(CH_3)_2 \xrightarrow[H_2SO_4]{CrO_3} (CH_3)_3CCOCH_3$ (among other products)

(c) $(CH_3)_2C{=}C(CH_3)_2 \xrightarrow{O_3}$ ozonide: H_3C, H_3C / C–O–C / O–O / CH_3, CH_3

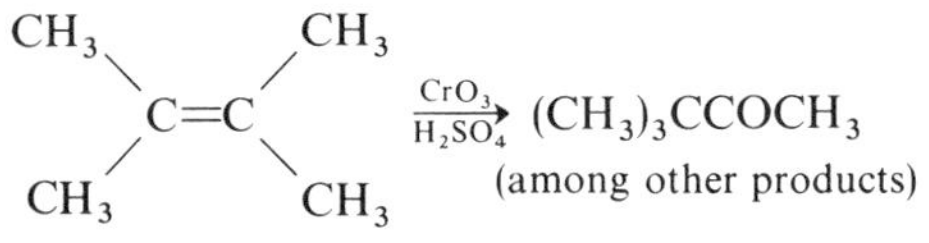

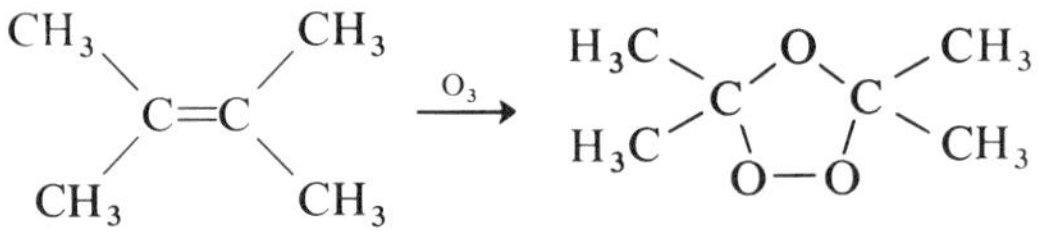

(d) 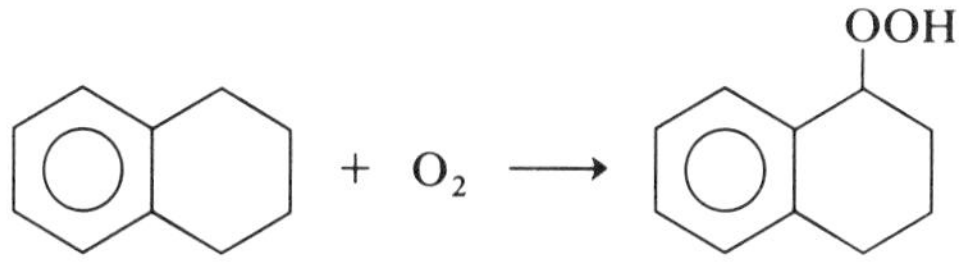 [tetralin] + O_2 ⟶ [1-tetralyl hydroperoxide, OOH]

2. Suggest reaction conditions (reagents, catalysts, solvents, etc.) suitable for effecting the following conversions. More than one step may be required.

(a) 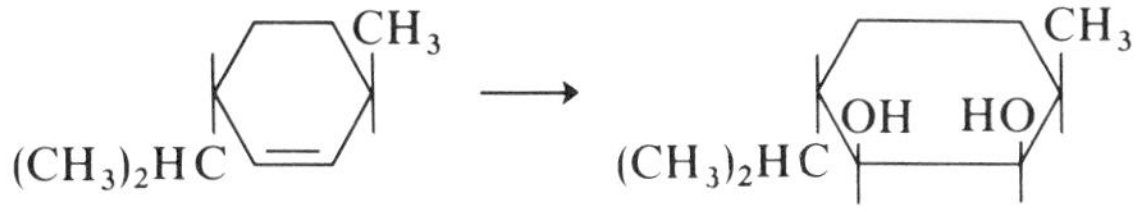 [$(CH_3)_2HC$, CH_3 cyclohexene] ⟶ [$(CH_3)_2HC$, CH_3, OH, HO cyclohexanediol]

(b) C_6H_5–$CH_2CH_2CH_3$ ⟶ $HOOCCH_2CH_2CH_3$

(c) H_3C–C_6H_4–CH_3 ⟶ HOOC–C_6H_4–COOH

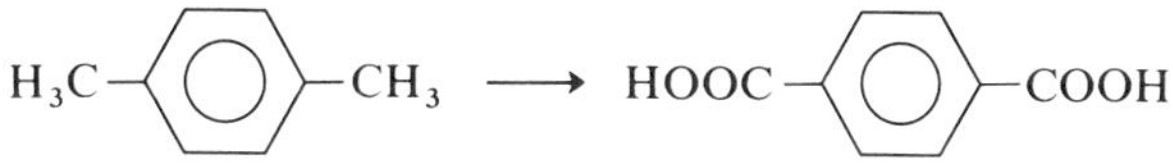

(d) 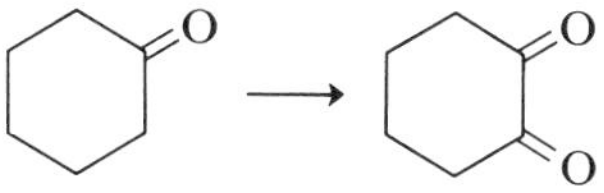 [cyclohexanone] ⟶ [cyclohexane-1,2-dione]

(e) $CH_3(CH_2)_7C(H)=C(H)(CH_2)_7CH_3$ ⟶ $CH_3(CH_2)_7C(=O)-CH(OH)(CH_2)_7CH_3$

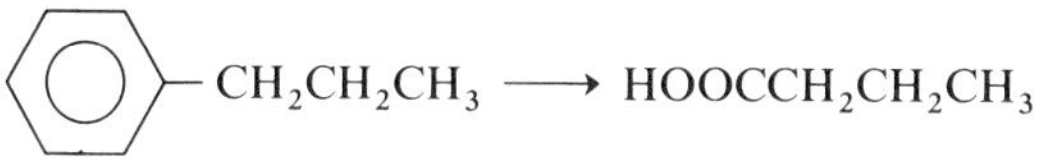

(f) 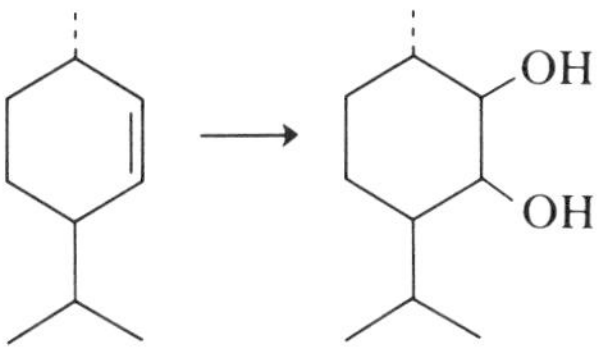 [4-isopropyl-1-methylcyclohexene] ⟶ [diol, OH, OH]

(g) 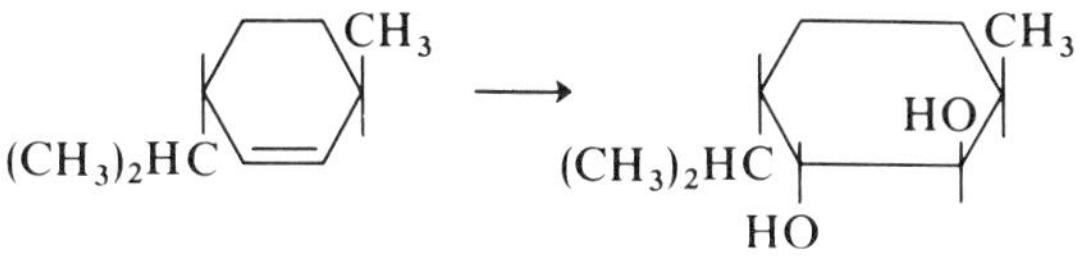 [$(CH_3)_2HC$, CH_3 cyclohexene] ⟶ [$(CH_3)_2HC$, CH_3, HO, HO cyclohexanediol]

(h) 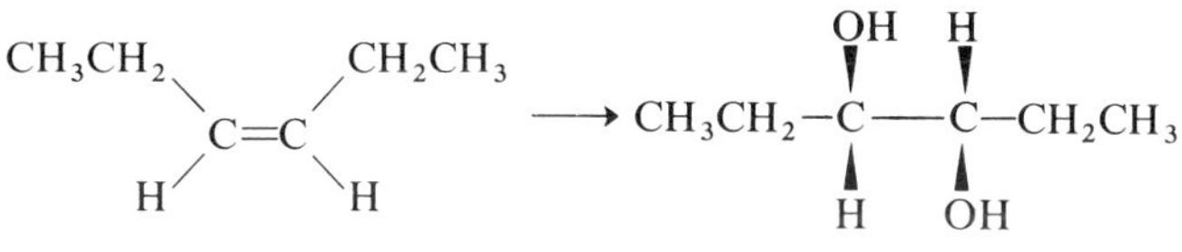 $CH_3CH_2(H)C=C(H)CH_2CH_3$ ⟶ $CH_3CH_2-C(OH)(H)-C(H)(OH)-CH_2CH_3$

racemic

3. Indicate all the expected products of the following reactions. Show stereochemistry where appropriate.

(a) 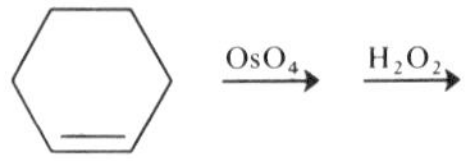 cyclohexene $\xrightarrow{OsO_4}$ $\xrightarrow{H_2O_2}$

(b) 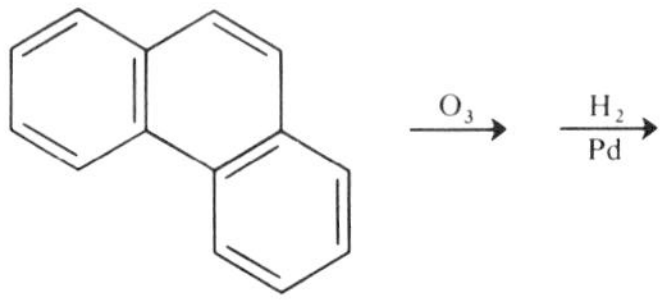 phenanthrene $\xrightarrow{O_3}$ $\xrightarrow[Pd]{H_2}$

(c) 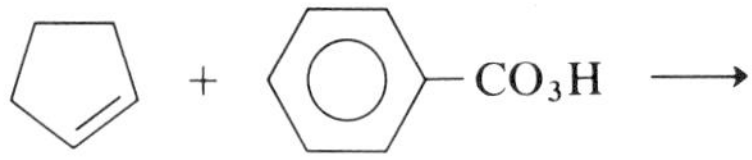 cyclopentene + C_6H_5—CO_3H $\longrightarrow$

4. Show by a three-dimensional drawing that cis addition of permanganate to oleic acid leads to the *erythro* diol shown on page 37.

5. Write balanced equations for the oxidation steps in Problems 1(a), 2(a), 2(c), 2(d), and 2(e).

4

Reductive Cleavage in Oxygen-Containing Functional Groups

Chapter 3 discussed procedures and reagents for introducing oxygen into hydrocarbon units of organic molecules. The present chapter deals with the reverse process—methods for removing oxygen and converting oxygen-containing groups to hydrocarbon units.

CONVERSION OF ALCOHOLS TO SATURATED HYDROCARBONS

Cleavage of a bond in the presence of hydrogen is called hydrogenolysis, and this reaction could provide a simple method for converting alcohols to saturated hydrocarbons. Unfortunately, it is usually quite difficult to hydrogenolyze the C—O bond in an alcohol, and there is no good general method for carrying out this reaction, although benzylic and, sometimes, allylic alcohols can be hydrogenolyzed, as we shall see in a later section of this chapter.* In fact, it is sufficiently difficult to hydrogenolyze an ordinary alcohol that treatment of 1-octadecanol with hydrogen at high pressure gives heptadecane, thus breaking a C—C bond as well as a C—O bond.

$$CH_3(CH_2)_{15}CH_2CH_2OH + 2H_2 \xrightarrow[200\text{ atm},\ 250^\circ]{\text{Ni(R)}} 90\%\ CH_3(CH_2)_{15}CH_3 + CH_4 + H_2O$$

To effect the conversion of an alcohol to a hydrocarbon, one must usually resign oneself to a two-step reaction sequence. A simple method is to

* The hydroxyl group in α-hydroxy ketones can sometimes be removed chemically, by Clemmensen reduction, as we shall see on page 62.

dehydrate the alcohol to give an alkene (employing, for example, one of the methods discussed by Saunders),* and then to hydrogenate the alkene over one of the catalysts described in Chapter 2. This method is relatively simple but can lead to rearrangements in the dehydration step and must lead to loss of configuration at an asymmetric adjacent carbon bearing the hydrogen lost.

$$-\underset{H}{\overset{|}{C}}-\underset{OH}{\overset{|}{C}}- \longrightarrow \;>C{=}C< \longrightarrow -\overset{|}{C}H\overset{|}{C}H-$$

Another two-step procedure for effecting the ROH ⟶ RH conversion is to convert the alcohol to the corresponding toluenesulfonate, then to treat the tosylate ester with lithium aluminum hydride, which replaces the tosylate by a hydrogen atom. The sequence works best with primary alcohols, somewhat less well with secondary alcohols, and not at all with tertiary alcohols. Since this is an S_N2 reaction, it proceeds with inversion of configuration, as has been shown for at least one secondary tosylate, by displacement of the tosylate group by deuterium with lithium aluminum deuteride.

$$C_6H_5-\underset{H}{\overset{OH}{C}}-CH_3 \xrightarrow{H_3C-C_6H_4-SO_2Cl} C_6H_5-\underset{H}{\overset{OSO_2-C_6H_4-CH_3}{C}}-CH_3 \xrightarrow{LiAlD_4} C_6H_5-\underset{D}{\overset{H}{C}}-CH_3$$

A third method for effecting the replacement of hydroxyl by hydrogen is to convert the alcohol to an alkyl halide and then to reduce off the halogen atom. The relative ease of replacement of halide is iodide > bromide > chloride. One method of removing the halide resembles the reduction of tosylates described, in that it involves treatment with lithium aluminum hydride. As with the tosylates, this method works best for primary, less well for secondary, and not at all for tertiary halides.

$$CH_3(CH_2)_2CH_2OH \xrightarrow{HBr} CH_3(CH_2)_3Br \xrightarrow{LiAlH_4} CH_3(CH_2)_2CH_3$$

Hydrogenolysis of the halide offers an attractive alternative to lithium aluminum hydride reduction in that this reaction proceeds most readily for tertiary halides and least readily for primary halides. For simple primary

* W. H. Saunders, *Ionic Organic Reactions*, in Foundations of Modern Organic Chemistry Series, Prentice-Hall, Inc., Englewood Cliffs, N. J., 1965.

and secondary halides, the iodide is the only useful halide. However, for tertiary halides, bromides or chlorides may be used as well. The choice of catalyst is usually palladium, but nickel may also be used. Allylic halides are especially readily reduced, and it is often possible to hydrogenolyze the halogen without reducing the adjacent carbon-carbon double bond, as we shall see later in this chapter.

$$(CH_3)_3C{-}OH \xrightarrow{HBr} (CH_3)_3CBr + H_2 \xrightarrow[\text{KOH, } CH_3OH,\ 1\text{ atm, r.t.}]{Pd/C} (CH_3)_3CH + HBr$$

An interesting aspect of halide hydrogenolysis is that this can be a particularly effective method for sparing cyclopropane rings, since they are not opened under the conditions required for halogen removal.

$$\text{1,1-dibromocyclopropane-2,3-dicarboxylic acid (Br, Br; HOOC, COOH)} + 2H_2\ (1\text{ atm}) \xrightarrow[\text{r.t.}]{\text{Ni, KOH, } CH_3OH} 80\%\ \text{cyclopropane-1,2-dicarboxylic acid (HOOC, COOH)}$$

The example shown serves to illustrate the point, although the dibromocyclopropane is not prepared from an alcohol but from dibromocarbene addition to maleic anhydride.

The last method we shall discuss for converting alcohols to hydrocarbons is oxidizing the alcohol to the corresponding aldehyde or ketone, then reducing the aldehyde or ketone to a hydrocarbon by one of the methods discussed in the section after the next.

$$>CHOH \longrightarrow >C{=}O \longrightarrow >CH_2$$

REDUCTION OF EPOXIDES

An additional example of reductive C—O bond cleavage is the opening of an epoxide ring. The reduction does not proceed all the way to a saturated hydrocarbon, but usually gives alcohols. The reduction can be effected by hydrogenation. Notice in the example shown that the two five-membered rings, which are not strained, are not opened in the hydrogenation.

$$\text{epoxide (}CH_2{-}O{-}C{-}H\text{) on furanose ring with } OCH_3 \text{ and isopropylidene } (O{-}C(CH_3)_2{-}O) + H_2\ (100\text{ atm}) \xrightarrow[CH_3OH,\ 65°]{Ni(R)} 95\%\ HO{-}C(CH_3){-}H \text{ on the same ring system}$$

Epoxides can also be opened reductively by lithium aluminum hydride. The direction of opening is usually predictable, since the reagent nearly always attacks at the less substituted end of an unsymmetrical epoxide.

$$H_3C-\overset{\overset{O}{\diagup\;\diagdown}}{HC-CH_2} \xrightarrow{LiAlH_4} CH_3\overset{\overset{OH}{|}}{C}HCH_3$$

REDUCTION OF ALDEHYDES AND KETONES TO HYDROCARBONS

Three useful methods effect the conversion of a carbonyl group to a methylene group: the Clemmensen reduction, the Wolff-Kishner reduction, and the hydrogenolysis of thioketals. The choice of one over another depends almost entirely upon what additional groups may be present in the molecule, whether they be sensitive to acid, base, or hydrogenation.

The Clemmensen reduction involves treatment of the carbonyl compound with a zinc amalgam in acid. The mechanism is somewhat obscure but apparently does not involve an intermediate alcohol, since, under the conditions of the reaction, the alcohol corresponding to the carbonyl compound is usually not reduced. An alkylzinc intermediate is a likely possibility, because it would be hydrolyzed by acid to give the hydrocarbon.

$$CH_3(CH_2)_5COCH_3 \xrightarrow[\text{HOAc, H}_2\text{O; HCl, reflux}]{\text{Zn-Hg}} 62\%\ CH_3(CH_2)_6CH_3$$

$$R-\overset{\overset{O}{||}}{C}-R + 2H^+ + 2Zn \longrightarrow R-\underset{\underset{Zn}{||}}{C}-R + H_2O + Zn^{2+}$$

$$\downarrow 2H^+$$

$$R-CH_2-R + Zn^{2+}$$

In Clemmensen reductions of α-hydroxy ketones, a two-electron reduction sometimes takes place to give the unsubstituted ketone.

$$(CH_2)_8\begin{bmatrix} CO \\ | \\ CHOH \end{bmatrix} + 2H^+ + Zn \longrightarrow 78\%\ (CH_2)_8\begin{bmatrix} CO \\ | \\ CH_2 \end{bmatrix} + H_2O + Zn^{2+}$$

The Wolff-Kishner reduction, on the other hand, takes place in a basic solution of potassium hydroxide in ethylene glycol or, better, triethylene glycol; the reducing agent is hydrazine. The reasonable mechanism for this reduction involves tautomerization of the intermediate hydrazone and subsequent loss of nitrogen during the attack by a second mole of hydroxide ion.

$$\mathrm{HOOC(CH_2)_4CO(CH_2)_4COOH} \xrightarrow[\mathrm{H_2NNH_2},\ \mathrm{HO(CH_2CH_2O)_3H}]{\mathrm{KOH}} 93\%\ \mathrm{HOOC(CH_2)_9COOH}$$

$$\mathrm{R{-}\overset{O}{\overset{\|}{C}}{-}R} + \mathrm{H_2NNH_2} \longrightarrow \mathrm{R{-}\overset{NNH_2}{\overset{\|}{C}}{-}R} \xrightarrow{\mathrm{OH^-}} \mathrm{R{-}\overset{N{=}NH^{\delta-}}{\overset{\|}{C}}{-}R}\ (\delta^-) \xrightarrow{\mathrm{ROH}} \mathrm{R{-}\overset{N{=}N{-}H}{\overset{|}{C}H}{-}R}$$

$$\xrightarrow{\mathrm{OH^-}} \mathrm{H_2O} + \mathrm{N_2} + \mathrm{R{-}\bar{C}H{-}R} \xrightarrow{\mathrm{ROH}} \mathrm{R{-}CH_2{-}R}$$

On attempted reduction of α-diketones, an alternative reaction takes place, presumably by the mechanism shown, to form a triple bond joining the carbons that were formerly involved in the carbonyl groups.

$$\mathrm{R{-}\overset{O}{\overset{\|}{C}}{-}\overset{O}{\overset{\|}{C}}{-}R} \xrightarrow[\text{Kishner}]{\text{Wolff-}} \mathrm{R{-}C{\equiv}C{-}R}$$

$$\mathrm{R{-}\overset{N{-}NH_2}{\overset{\|}{C}}{-}\overset{O}{\overset{\|}{C}}{-}R} \xrightarrow{\mathrm{OH^-}} \mathrm{R{-}\overset{N{=}N{-}H}{\overset{|}{C}}{=}\overset{OH}{\overset{|}{C}}{-}R} \xrightarrow{\mathrm{OH^-}} \mathrm{R{-}C{\equiv}C{-}R} + \mathrm{N_2} + \mathrm{H_2O}$$

Aldehydes and ketones may be readily converted with ethanedithiol or ethyl mercaptan to the corresponding thioacetals and thioketals, which display the propensity of sulfur-containing compounds toward desulfurization noted first for thiophene in Chapter 2. Replacement of the sulfur atoms by hydrogens then gives saturated hydrocarbons. In the example, 1 mole of ethanedithiol was used to give the intermediate monothioketal shown. Subsequent reduction with nickel takes place only at the thioketal and not at the carbonyl group.

$$\mathrm{OC\langle {(CH_2)_7 \atop (CH_2)_7} \rangle CO} + \mathrm{HSCH_2CH_2SH} \longrightarrow \mathrm{(CH_2S)_2C\langle {(CH_2)_7 \atop (CH_2)_7} \rangle CO} \xrightarrow{\mathrm{Ni}} \mathrm{\ulcorner(CH_2)_{15}{-}CO\urcorner}$$

REDUCTION OF ACIDS AND THEIR DERIVATIVES TO HYDROCARBONS

Apparently, no selective general method exists for converting an acid, or for that matter an ester, acid chloride, or anhydride, directly to a methyl group. In nearly every case, it is necessary to reduce the acid in two stages—first to the alcohol or aldehyde, and thence to the methyl group by one of the reactions in the preceding two sections.

REDUCTION OF BENZYLIC AND ALLYLIC COMPOUNDS TO HYDROCARBONS

We observed in Chapter 3 that the benzylic position is especially susceptible to attack by oxygen. The benzylic position is also quite susceptible to catalytic hydrogenolysis in which a substituent is replaced by a hydrogen atom. This is true of benzyl alcohols and also of benzyl ethers and benzyl esters, all of which give alkyl benzenes.

$$C_6H_5-CH(OH)CH(NHCH_3)CH_3 + H_2\ (3\ \text{atm}) \xrightarrow[60°,\ H_2SO_4]{Pd/C} C_6H_5-CH_2CH(NHCH_3)CH_3$$

$$C_6H_5-CH_2O-C_6H_5 + H_2 \xrightarrow{Ni} C_6H_5-CH_3 + HO-C_6H_5$$

Because of their susceptibility to hydrogenolysis, benzyl esters have been put to considerable use as protective groups in peptide chemistry. Carbobenzoxy chloride (benzyl chloroformate) reacts with an amine to give its benzyloxycarbonyl derivative. The derivative is an amide, which is neutral and therefore unreactive toward many reagents, but the free amine is regenerated easily by catalytic hydrogenolysis of the benzyl group.

$$HO-C_6H_4-CH_2CH(NH_2)CONHR \xrightarrow{C_6H_5-CH_2OCOCl} HO-C_6H_4-CH_2CH(NHCOOCH_2-C_6H_5)CONHR$$

$$HO-C_6H_4-CH_2CH(NHCOOCH_2-C_6H_5)CONHR \xrightarrow[HCl,\ EtOH,\ r.t.]{H_2,\ Pd/C} HO-C_6H_4-CH_2CH(NH_2)CONHR + C_6H_5-CH_3 + CO_2$$

Carbonyl groups in the benzylic position are also highly susceptible to hydrogenolysis. In the example shown, the conditions are sufficiently mild that the carbonyl group adjacent to the benzene ring is hydrogenolyzed quantitatively, whereas the carbonyl group one carbon atom removed from the benzene ring is not attacked.

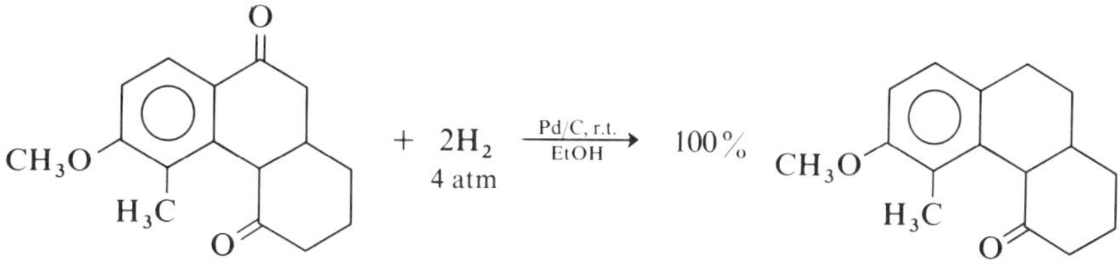

CH_3O H_3C O O + $2H_2$ 4 atm $\xrightarrow[\text{EtOH}]{\text{Pd/C, r.t.}}$ 100% CH_3O H_3C O

Benzylic amines can also be hydrogenolyzed to alkyl benzenes, as can benzyl mercaptans and benzyl halides. Allylic sulfur- and halogen-containing compounds also undergo hydrogenolysis very readily—sufficiently so that the adjacent carbon-carbon double bonds are not hydrogenated under the conditions required. The example shows replacement of an allylic bromide by deuterium.

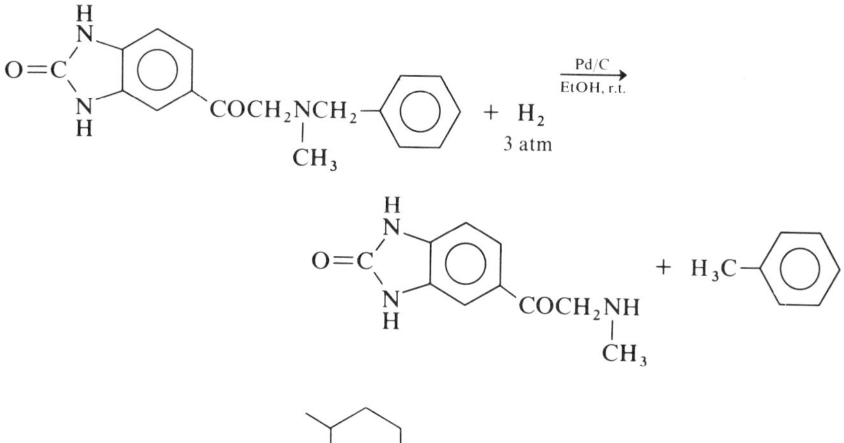

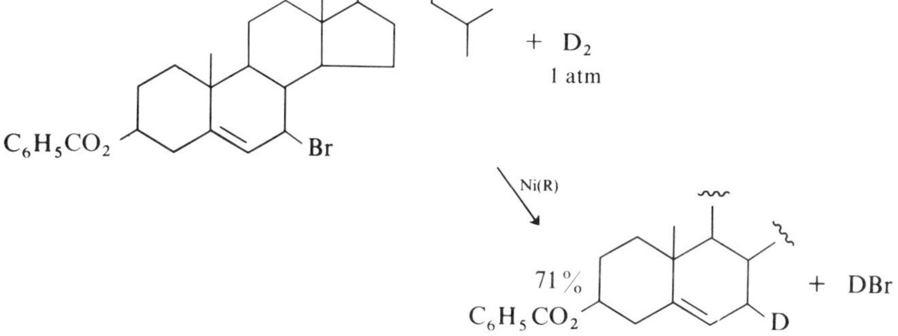

H N O=C N H $COCH_2NCH_2$ CH_3 + H_2 3 atm $\xrightarrow[\text{EtOH, r.t.}]{\text{Pd/C}}$

H N O=C N H $COCH_2NH$ CH_3 + H_3C

$C_6H_5CO_2$ Br + D_2 1 atm

Ni(R)

71% $C_6H_5CO_2$ D + DBr

REFERENCES

Augustine (*Reduction*): Chap. 1, M. N. Rerick, "The Chemistry of the Mixed Hydrides."

Augustine (*Reduction*): Chap. 3, W. Reusch, "Deoxygenation of Carbonyl Compounds."

Hartung, W. H., "Hydrogenolysis of Benzyl Groups Attached to Oxygen, Nitrogen, or Sulfur," *Org. Reactions*, **7**, 263 (1953).

House: Chap. 2, "Metal Hydride Reductions and Related Reactions."

Martin, E. L., "The Clemmensen Reduction," *Org. Reactions*, **1**, 155 (1942).

Rylander, P. N., "Catalytic Hydrogenation over Platinum Metals," Academic Press, Inc., New York, 1967, Part 6 ("Hydrogenolysis").

Staschewski, D., "The Mechanism of the Clemmensen Reduction," *Angew. Chem.*, **71**, 726 (1959).

Todd, D., "The Wolff-Kishner Reduction," *Org. Reactions*, **4**, 378 (1948).

Wheeler, O. H., "Reduction of Carbonyl Groups," in *Chemistry of the Carbonyl Group*, S. Patai, ed., John Wiley & Sons, Inc., New York, 1966, Chap. 11.

PROBLEMS

1. Write a detailed mechanism for each of the following reactions, including all intermediates and/or transition states.

(a) 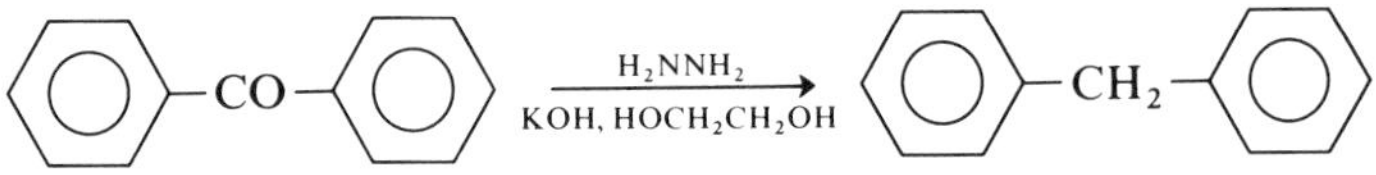 $C_6H_5\text{—CO—}C_6H_5 \xrightarrow[\text{KOH, HOCH}_2\text{CH}_2\text{OH}]{H_2NNH_2} C_6H_5\text{—CH}_2\text{—}C_6H_5$

(b) $CH_3CH_2\overset{O}{\overset{\|}{C}}\text{—}\overset{O}{\overset{\|}{C}}CH_2CH_3 \xrightarrow[\text{KOH, }\Delta]{H_2NNH_2} CH_3CH_2C{\equiv}CCH_2CH_3$

2. Suggest reaction conditions (reagents, catalysts, solvents, etc.) suitable for effecting the following conversions. More than one step may be required.

(a) 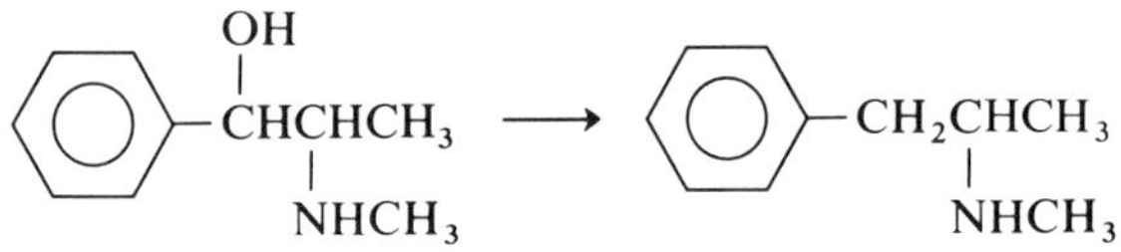 $C_6H_5\text{—CH(OH)CH(NHCH}_3\text{)CH}_3 \longrightarrow C_6H_5\text{—CH}_2\text{CH(NHCH}_3\text{)CH}_3$

(b) 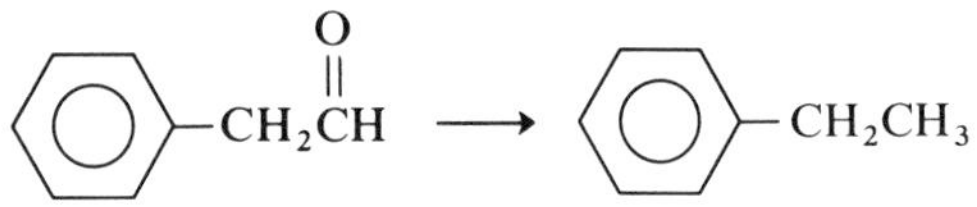 $C_6H_5\text{—CH}_2\overset{O}{\overset{\|}{C}}H \longrightarrow C_6H_5\text{—CH}_2CH_3$

(c) 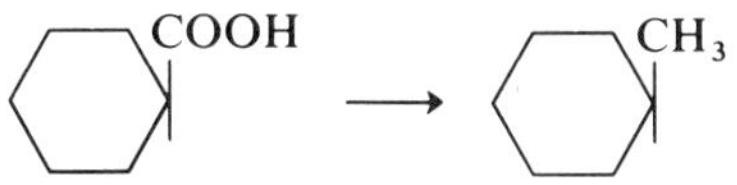 Cyclohexane bearing COOH and a methyl-type substituent → cyclohexane bearing CH_3

(d) 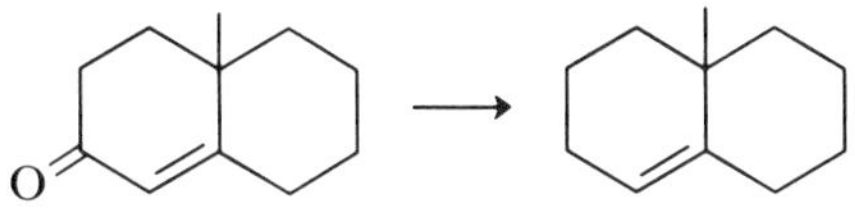 Bicyclic enone (O=) → bicyclic alkene

(e) $C_6H_5CH_2OCONHCH(CH_2C_6H_5)CONHCH(CH_3)COOH \longrightarrow H_2NCH(CH_2C_6H_5)CONHCH(CH_3)COOH$

(f) 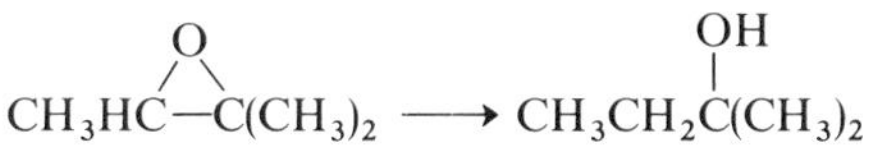 $CH_3HC\overset{O}{—}C(CH_3)_2 \longrightarrow CH_3CH_2C(OH)(CH_3)_2$

(g) $C_6H_5CH_2CH_2CH_2OH \longrightarrow C_6H_5CH_2CH_2CH_3$

(h) 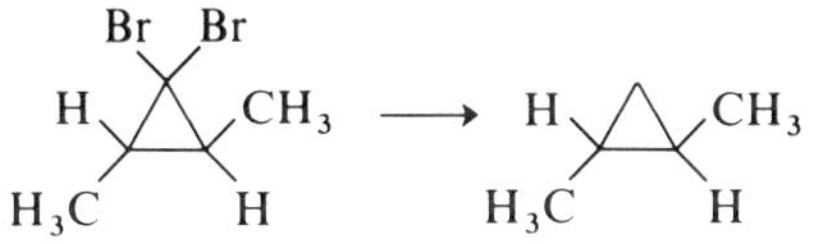 Br, Br; H, CH_3; H_3C, H (dibromocyclopropane) $\longrightarrow$ H, CH_3; H_3C, H (cyclopropane)

3. Indicate the expected product of the following reaction.

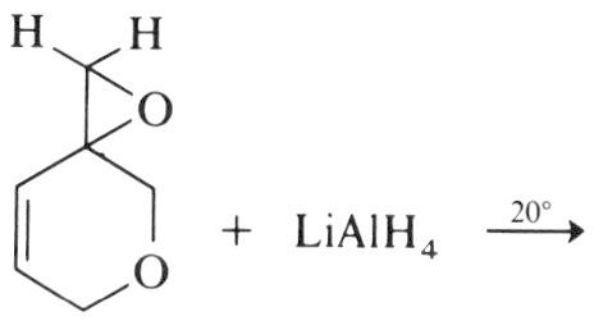

H, H; O; O $\quad + \ LiAlH_4 \xrightarrow{20°}$

4. Write balanced equations for the reduction steps shown in Problems 1(a), 2(a), 2(d), and 2(f).

5

Oxidation of Alcohols and Phenols

The oxidation of alcohols is the principal subject of this chapter, although we shall deal briefly with oxidations of phenols and ethers. In Chapter 3, we saw that it is often difficult to isolate alcohols as products of oxidation, for they themselves tend to be rather susceptible to further oxidation. This was true, for example, in the permanganate oxidation of alkenes on pages 37 and 38, which gave hydroxy ketones from oleic acid and a dialdehyde from norbornene in addition to the vicinal glycols. It was also true in the chromic oxide oxidation of 2,3-dimethyl-2-butene on page 44, which gave acetone, pinacolone, and pivalic acid as products rather than the vicinal glycol. Permanganate, chromic oxide and many other reagents are, thus, capable of oxidizing alcohols.

$$(CH_3)_2C{=}C(CH_3)_2 \longrightarrow \left[(H_3C)_2C(OH){-}C(OH)(CH_3)_2\right] \longrightarrow (CH_3)_2C{=}O \quad O{=}C(CH_3)_2$$

CHROMIC ACID OXIDATION

The reagent most commonly used in the laboratory to effect the oxidation of an alcohol is chromic acid, a somewhat generic term used to describe the various states of hydration of Cr^{VI}, among them $HCrO_4^-$, $Cr_2O_7^{2-}$, and CrO_3. Not only is this the most common reagent employed but its mechanism of oxidation has been the most extensively studied.

The usual product from the oxidation of a secondary alcohol is a ketone; primary alcohols can give either aldehydes, acids, or esters; isolated tertiary alcohols are not normally oxidized by dichromate. Conversion of a secondary alcohol to a ketone is easy, although the conditions must be controlled to prevent overoxidation to acids.

$$CH_3CH_2CH(OH)(CH_2)_2CH_3 \xrightarrow[< 30^\circ]{CrO_3,\ HOAc} 63\%\ CH_3CH_2C(=O)CH_2CH_2CH_3$$

$$\text{cyclohexanol} \xrightarrow[H_2SO_4,\ 55^\circ]{K_2Cr_2O_7} \text{cyclohexanone}$$

$$\text{cyclohexanol} \xrightarrow[H_2SO_4,\ > 60^\circ]{K_2Cr_2O_7} HOOC(CH_2)_4COOH$$

Conversion of a primary alcohol to an aldehyde requires very careful control of conditions, because the aldehyde itself is susceptible to further oxidation. A procedure often employed with low molecular weight alcohols involves use of a limited amount of oxidant (dropwise addition) and distillation of the aldehyde, lower boiling than the alcohol, as it is formed, to remove it from the oxidant. Acids are the products of oxidation of the aldehydes (see Chapter 7). The formation of an ester probably results from addition of a second mole of alcohol to the aldehyde, followed by the oxidation of a hemiacetal to the ester.

$$CH_3CH_2CH_2OH \xrightarrow[\text{added dropwise, distill}]{K_2Cr_2O_7,\ \text{dil}\ H_2SO_4} 49\%\ CH_3CH_2CH(=O)$$

$$CH_3CH_2CH_2OH \xrightarrow[H_2SO_4,\ 115^\circ]{K_2Cr_2O_7} 65\%\ CH_3CH_2COOH$$

$$CH_3(CH_2)_2CH_2OH \xrightarrow[H_2SO_4,\ < 20^\circ]{Na_2Cr_2O_7} 47\%\ CH_3(CH_2)_2CH_2OCO(CH_2)_2CH_3$$

$$\searrow \quad CH_3(CH_2)_2{-}CH(OH)(OCH_2(CH_2)_2CH_3) \quad \nearrow$$

One of the attractive features of chromic acid oxidation is the many forms in which the oxidant can be employed. Some of these can effect oxidation under conditions mild enough that other oxidizable functional groups—such as carbon-carbon double bonds or triple bonds—are not attacked. For example, a secondary alcohol can be oxidized in the presence of a carbon-carbon triple bond by chromic oxide in acetone (the so-called Jones reagent), and cinnamyl alcohol can be oxidized to cinnamaldehyde by a chromic oxide-pyridine complex (Sarett's reagent). Some of these specific conditions have been developed in connection with studies on the chemistry of steroids.

$$CH_3CH(OH)C{\equiv}C(CH_2)_3CH_3 \xrightarrow[CH_3COCH_3,\ 5\text{–}10°]{CrO_3,\ \text{dil}\ H_2SO_4} 80\%\ CH_3COC{\equiv}C(CH_2)_3CH_3$$

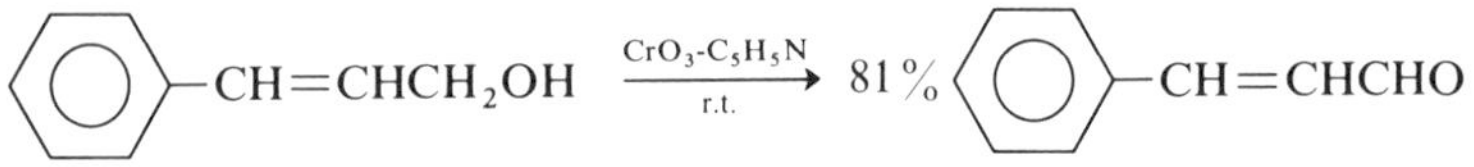

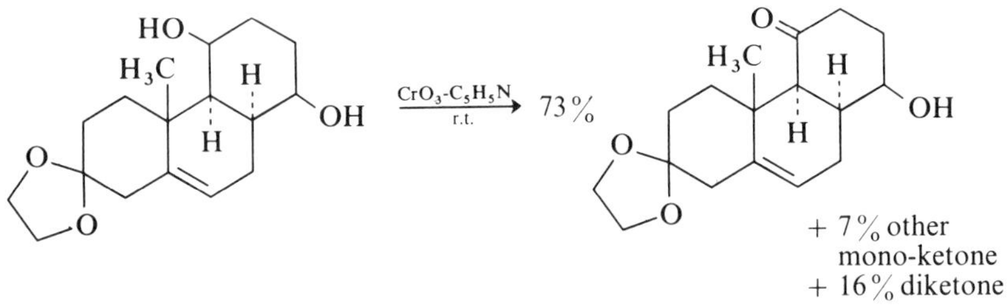

$$C_6H_5\text{–}CH{=}CHCH_2OH \xrightarrow[\text{r.t.}]{CrO_3\text{-}C_5H_5N} 81\%\ C_6H_5\text{–}CH{=}CHCHO$$

$$\xrightarrow[\text{r.t.}]{CrO_3\text{-}C_5H_5N} 73\%$$

HO, H_3C, H, H, OH, O, O — O, H_3C, H, H, OH, O, O

+ 7% other mono-ketone
+ 16% diketone

Much effort has been directed toward elucidation of the mechanism of oxidation by chromic oxide, notably by Westheimer. Some of the results of these efforts are found in the following observations. First, there are apparently two competing, related oxidation mechanisms, as indicated by the two-term rate equation shown for isopropyl alcohol.

$$\text{rate} = k_1[HCrO_4^-][(CH_3)_2CHOH][H^+] + k_2[HCrO_4^-][(CH_3)_2CHOH][H^+]^2$$

Second, a sizable deuterium isotope effect is shown for substitution of the carbinol hydrogen (the hydrogen attached to the carbon bearing the oxygen atom) by deuterium in isopropyl alcohol.

$$\frac{k_H}{k_D} = 7 \quad \text{for } CH_3\text{–}C(OH)(D)\text{–}CH_3 \text{ vs. } CH_3\text{–}C(OH)(H)\text{–}CH_3$$

Third, the intermediacy of a Cr^{IV} species can be demonstrated by trapping of that species with Mn^{II}.

$$R_2CHOH + Cr^{VI} \longrightarrow R_2CO + Cr^{IV} + 2H^+$$
$$Cr^{IV} + Cr^{VI} \longrightarrow 2Cr^{V}$$
$$\text{Trapped:} \quad Cr^{IV} + Mn^{II} \longrightarrow Cr^{III} + Mn^{III} \longrightarrow MnO_2 + Mn^{II}$$
$$Cr^{V} + R_2CHOH \longrightarrow Cr^{III} + R_2CO + 2H^+$$

Fourth, variation of the rate of oxidation of substituted α-phenylethyl alcohols to substituted acetophenones gives a negative value (–1.00) for ρ

in the Hammet equation,* indicating some electron deficiency at the α carbon in the transition state.

$$X\text{-}C_6H_4\text{-}CH(OH)CH_3 \longrightarrow X\text{-}C_6H_4\text{-}COCH_3$$

$$\rho = -1.01$$

A mechanism that fits these data is the one shown, involving the formation of a chromate ester of the alcohol and its decomposition by two pathways. The rate-determining step must be the decomposition of the chromate ester, as indicated by the deuterium isotope effect at the carbinol hydrogen. A base is required to accept the carbinol hydrogen. In most cases in aqueous solution, the base is presumably water, although an intramolecular reaction cannot be excluded.

$$H_2O \rightarrow H\text{-}CR_2\text{-}O\text{-}Cr(=O)_2OH \longrightarrow Cr^{IV} \quad \text{or} \quad H\text{-}CR_2\text{-}O\text{-}Cr(=O)_2OH \longrightarrow Cr^{IV}$$

$$R_2CHOH + HCrO_4^- + H^+ \longrightarrow R_2CHO{-}CrO_3H + H_2O \xrightarrow{k_1} R_2CO + Cr^{IV}$$

$$R_2CHOCrO_3H + H^+ \longrightarrow R_2CHOCrO_3H_2^+ \xrightarrow{k_2} R_2CO + Cr^{IV}$$

Further support for this mechanism is provided by the isolation of a diisopropyl chromate ester, which in pyridine decomposes to acetone.

$$(CH_3)_2CH{-}O{-}CrO_2{-}O{-}CH(CH_3)_2 \xrightarrow{\text{pyridine}} (CH_3)_2C{=}O + {}^-O_2CrO{-}CH(CH_3)_2 + C_5H_5N{\cdot}H^+$$

can be prepared

Further support of an intermediate chromate ester is provided by the results of oxidation of conformationally rigid secondary alcohols, in which the more hindered alcohol (usually axial) is oxidized more rapidly. This is usually interpreted in terms of relief of steric compression of the ground state of the chromate ester on reaching the transition state, although greater

* The Hammett equation is discussed in detail in L. M. Stock, *Aromatic Substitution Reactions*, in Foundations of Modern Organic Chemistry Series, Prentice-Hall, Inc., Englewood Cliffs, N. J., 1968.

accessibility of a less hindered (usually equatorial) carbinol hydrogen by the base is perhaps a contributing factor.

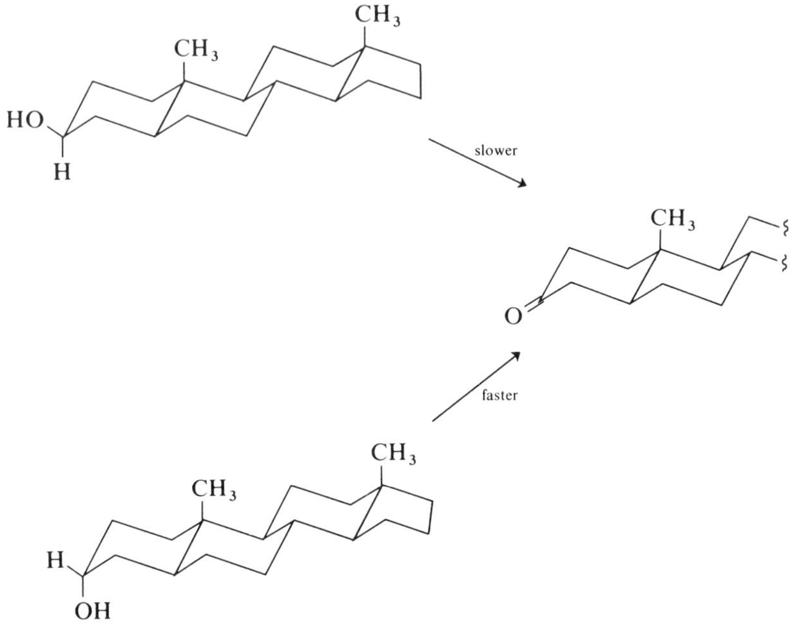

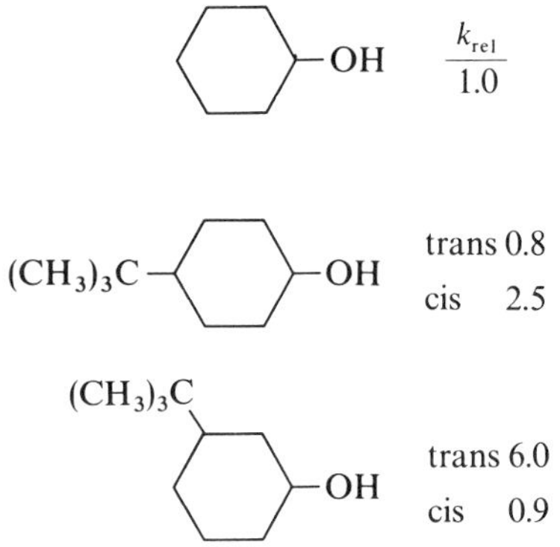

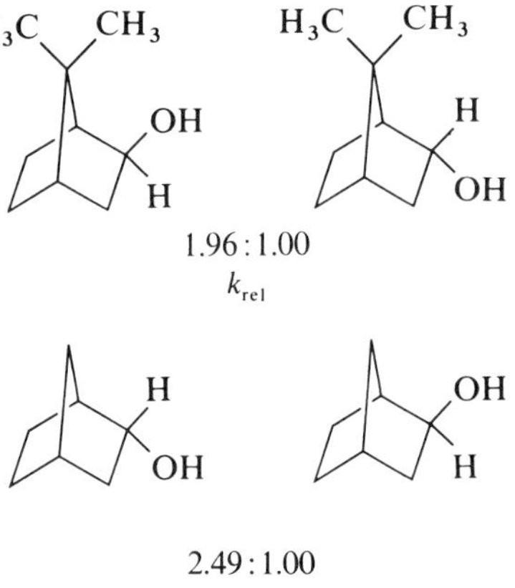

CH_3 CH_3 HO H — slower →

CH_3 O

CH_3 CH_3 H OH — faster →

	k_{rel}
cyclohexanol —OH	1.0
$(CH_3)_3C$—cyclohexyl—OH (4-substituted)	trans 0.8, cis 2.5
$(CH_3)_3C$—cyclohexyl—OH (3-substituted)	trans 6.0, cis 0.9

H_3C CH_3 OH H H_3C CH_3 H OH

1.96 : 1.00
k_{rel}

H OH   OH H

2.49 : 1.00

PERMANGANATE OXIDATION

Oxidation of alcohols by potassium permanganate is base catalyzed, and the principal decision on whether to oxidize an alcohol with chromic acid or permanganate usually rests on whether the reaction is to be carried out in acid or base. Oxidation of secondary (2°) alcohols by permanganate frequently gives good yields of ketones, although overoxidation can be a problem where there are α-hydrogens, due to enol formation. Primary (1°) alcohols almost never give aldehydes with permanganate but are overoxidized either to carboxylic acids or, via the enols, to shorter chain compounds.

$$2^\circ \quad C_6H_5CH(OH)CH(CH_3)_2 \xrightarrow[\text{HOAc, }30^\circ]{KMnO_4} 71\% \; C_6H_5COCH(CH_3)_2$$

$$1^\circ \quad CH_3(CH_2)_3CH(C_2H_5)CH_2OH \xrightarrow[H_2O,\ 12\text{ hr},\ 25^\circ]{KMnO_4,\ NaOH} 74\% \; n\text{-}C_4H_9CH(C_2H_5)COOH$$

Overoxidation:

$$2^\circ \quad RCH_2CHOHR' \longrightarrow RCH_2COR' \xrightarrow{OH^-} RCH{=}C(OH){-}R' \longrightarrow RCOOH + HOOCR'$$

$$1^\circ \quad RCH_2CH_2OH \longrightarrow RCH_2CHO \longrightarrow RCH{=}CHOH \longrightarrow RCOOH + CO_2$$

$$RCH_2CHO \longrightarrow RCH_2COOH$$

Permanganate oxidation has been somewhat less studied than oxidation with chromate, although the mechanism seems reasonably clear. First, the rate is dependent on the first power of the concentration of base, which presumably is employed in forming an alkoxide ion.

$$\text{rate} = k[R_2CHOH][MnO_4^-][OH^-]$$

Second, an isotope effect is observed at the carbinol hydrogen, indicating that the rate-determining step (r.d. step) must involve loss of carbinol hydrogen.

$$\frac{k_H}{k_D} = 6.6 \quad \text{at} \quad 25^\circ \quad \text{for } (C_6H_5)_2CHOH \text{ vs. } (C_6H_5)_2CDOH$$

There is, however, only a rather small substituent effect in the oxidation of phenyl-substituted carbinols. For the alcohol shown, at pH 13, $\rho \cong 0$. This

$$X\text{-}C_6H_4\text{-}CH(OH)CF_3$$

has been interpreted as suggesting a radical anion, formed by a one-electron oxidation, as the primary product rather than a neutral ketone, formed by a two-electron oxidation. The mechanism usually accepted is that which follows. Manganate ion (MnO_4^{2-} and $HMnO_4^-$) disproportionates to permanganate and manganese dioxide.

$$R_2CHOH \xrightarrow{OH^-} R_2CHO^-$$

$$R_2CHO^- + MnO_4^- \xrightarrow[\text{step}]{\text{r.d.}} R_2\dot{C}{-}O^- + HMnO_4^-$$

$$R_2\dot{C}{-}O^- \xrightarrow{MnO_4^-} R_2CO + MnO_4^{2-}$$

CERIC ION OXIDATION

Ceric ammonium nitrate is a reagent which has been recently introduced for the oxidation of alcohols. Although the competing reaction of oxidative cleavage can lead to reduced yields of ketones from secondary alcohols, the reagent can be very useful for the selective oxidation of primary alcohols to aldehydes.

$$\text{c-}C_3H_5{-}CH_2OH \xrightarrow[75^\circ]{(NH_4)_2Ce(NO_3)_6} 64\%\ \text{c-}C_3H_5{-}CHO$$

$$CH_3O{-}C_6H_4{-}CH_2OH \longrightarrow 80\%\ CH_3O{-}C_6H_4{-}CHO$$

$$CH_3O{-}C_6H_4{-}CH(OH)CH_2CH_3 \longrightarrow 18\%\ CH_3O{-}C_6H_4{-}C(=O)CH_2CH_3 + 60\%\ CH_3O{-}C_6H_4{-}CHO$$

NITRIC ACID OXIDATION

Nitric acid is very seldom used for selective oxidation of alcohols to ketones since other reagents are better adapted to that purpose. Nitric acid is a very vigorous oxidizing agent, however, and is employed commercially in the formation of adipic acid (used in Nylon manufacture) by the oxidation of cyclohexanol.

$$\text{cyclohexanol (OH)} \xrightarrow[(NH_4)_2V_2O_6]{50\%\ HNO_3,\ 60^\circ} HOOC(CH_2)_4COOH$$

CATALYTIC DEHYDROGENATION OF ALCOHOLS

Oxidation of a primary alcohol to an aldehyde or a secondary alcohol to a ketone is schematically at least nothing more than removal of hydrogen from a $-\underset{|}{\text{CHOH}}$ group, and methods have been found for effecting this removal on metal catalysts similar to those employed in the dehydrogenation of hydrocarbons. One of the most general catalytic oxidations of alcohols takes place over copper at elevated temperatures. The reaction is neither mild nor selective, however, and is usually employed only on an industrial scale. Similarly, hydrogen can be removed at high temperatures in the presence of a silver catalyst. These two reactions are complicated by the reverse process, with hydrogen reducing the carbonyl group back to the carbinol. This reverse process can be avoided either by sweeping the hydrogen out as it is formed or by adding enough oxygen to react with it to give water. The former procedure is used with copper, the latter with silver.

$$R_2CHOH \underset{300^\circ}{\overset{Cu}{\rightleftharpoons}} R_2CO + H_2$$

$$CH_3OH + \tfrac{1}{2}O_2 \xrightarrow{Ag,\ 250^\circ} CH_2O + H_2O$$

Dehydrogenation under considerably milder conditions takes place over the more active catalyst platinum. Oxygen is bubbled into the solution to remove the hydrogen, and the reactions can be carried out at room temperature. Considerable selectivity is observed as to which hydroxyl groups are oxidized in polyhydroxy compounds. In cyclitols and sugar derivatives, secondary alcohols with axial hydroxyl groups are most easily oxidized, primary alcohols next most easily, and equatorial hydroxyls most difficultly. In steroids, it is nearly always the relatively unhindered C-3 hydroxyl that is oxidized, although primary alcohols are occasionally attacked.

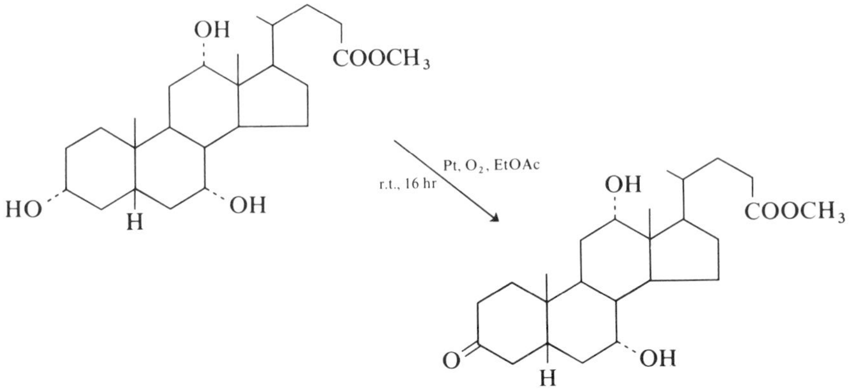
OH
COOCH3
HO
H
OH
Pt, O2, EtOAc
r.t., 16 hr
OH
COOCH3
O
H
OH

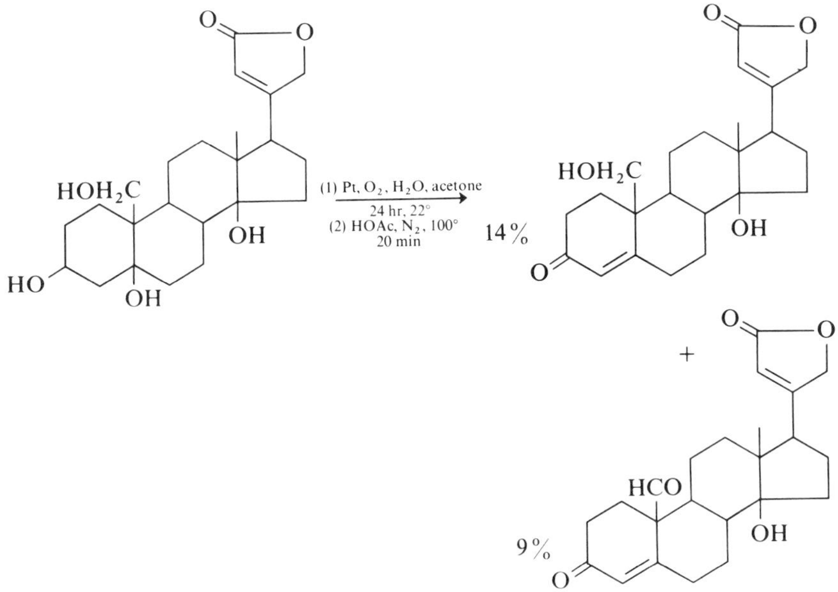
O
O
HOH2C
OH
HO
OH
(1) Pt, O2, H2O, acetone
24 hr, 22°
(2) HOAc, N2, 100°
20 min
14%
HOH2C
OH
O
+
HCO
OH
9%
O

OPPENAUER OXIDATION

Another equilibrium reaction that has been much used for the oxidation of secondary alcohols to ketones is the Oppenauer oxidation. In this reaction, a secondary alcohol (*A*H) and a ketone (*B*) are interconverted to another ketone (*A*) and the other secondary alcohol (*B*H) in the presence of a basic catalyst, usually an aluminum alkoxide such as aluminum isopropoxide. Because the reaction is reversible, it can be used to form either a ketone or a secondary alcohol, depending on which is desired. The reverse reaction, known as the Meerwein-Ponndorf-Verley reduction, will be discussed in

Chapter 6. A shift in one direction or the other is normally achieved by employing an excess of a cheap ketone (acetone or cyclohexanone) to effect oxidation or an excess of a cheap secondary alcohol (isopropyl alcohol) to effect reduction. The reaction can be further shifted toward oxidation by distilling out isopropyl alcohol as it is formed, along with excess acetone used as oxidant.

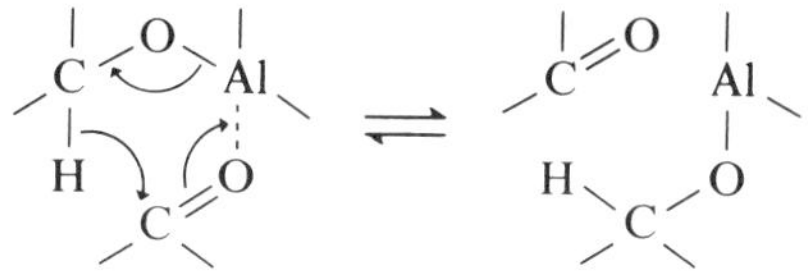

Equatorial hydroxyl groups (*e*) tend to be oxidized preferentially over axial hydroxyl groups (*a*), and the principal side reactions are those to be expected in basic solution, such as isomerization of unconjugated carbon-carbon double bonds into conjugation with the ketones formed.

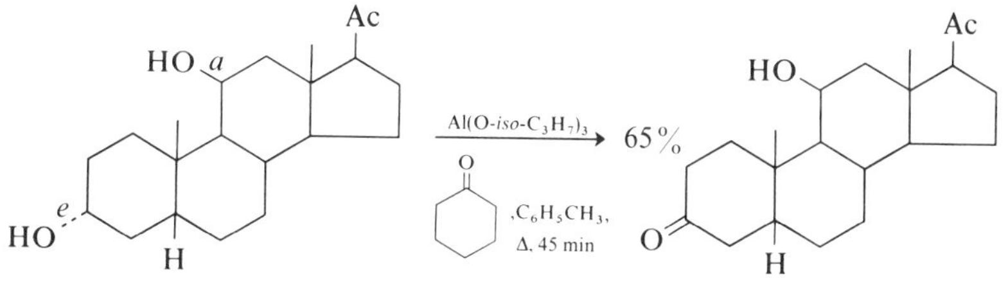

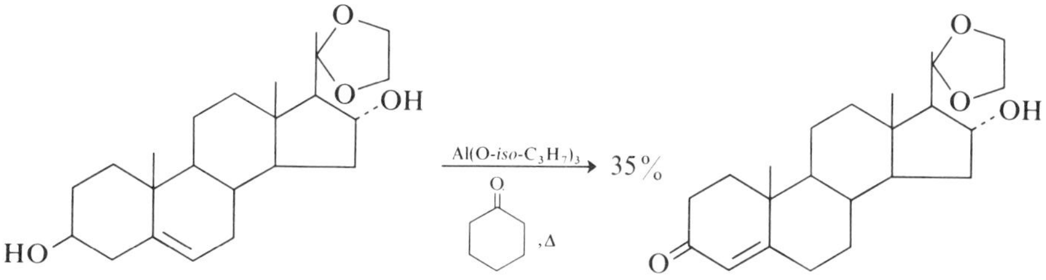

$Al(O\text{-}iso\text{-}C_3H_7)_3$, cyclohexanone, $C_6H_5CH_3$, Δ, 45 min — 65%

$Al(O\text{-}iso\text{-}C_3H_7)_3$, cyclohexanone, Δ — 35%

DIMETHYL SULFOXIDE OXIDATION

We have already discussed in this chapter a number of methods suitable for oxidizing secondary alcohols to ketones. Many fewer reactions are available for oxidizing primary alcohols to aldehydes, since the aldehydes themselves, as noted before, are susceptible to oxidation, giving mainly acids. A few special reactions that yield aldehydes have been noted but no general method. A very useful general method for the preparation of aldehydes involves displacement of a primary halide or tosylate or other alcohol derivative with dimethyl sulfoxide (DMSO). Ketones can be formed, too, but somewhat less readily. This method takes advantage of the preferential displacement of a primary group over a secondary group, so that, in fact, primary alcohols are more easily oxidized than secondary. An example is the formation of octanal.

$$CH_3(CH_2)_6CH_2OTs \xrightarrow[150°,\ 3\ min \atop DMSO]{NaHCO_3} 78\%\ CH_3(CH_2)_6CHO$$

In some reactions, the intermediate alcohol derivative is formed in situ for displacement. Intermediates in the two examples shown are the chloroformate, RCH_2OCOCl, and the phosphate, $CH_3(CH_2)_6CH_2OPO_3H_2$. Dicyclohexylcarbodiimide ($C_6H_{11}N{=}C{=}NC_6H_{11}$, abbreviated DCCD) serves to facilitate phosphate formation from the alcohol and phosphoric acid.

$$RCH_2OH \xrightarrow{COCl_2} \xrightarrow{DMSO} \xrightarrow{Et_3N} RCHO$$

$$CH_3(CH_2)_6CH_2OH \xrightarrow[DCCD \atop DMSO,\ r.t.]{H_3PO_4} 70\%\ CH_3(CH_2)_6CHO$$

Only those halides most reactive toward nucleophilic displacement can be used in the reaction. This includes all primary alkyl iodides as well as benzylic and allylic chlorides and bromides, and those adjacent to carbonyl groups.

$$RCH_2I \xrightarrow[DMSO]{NaHCO_3} RCHO$$

$$RCOCH_2X, RO_2CCH_2X, C_6H_5CH_2X \xrightarrow[DMSO]{NaHCO_3} RCOCHO, RO_2CCHO, C_6H_5CHO$$

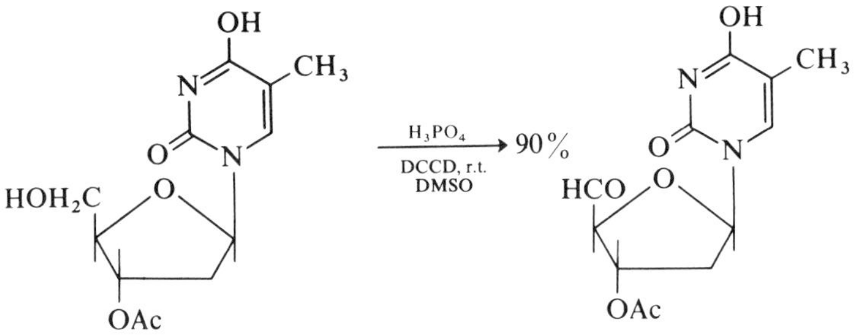

OH, CH₃, N, O, N, HOH₂C, O, OAc $\xrightarrow[DCCD,\ r.t. \atop DMSO]{H_3PO_4}$ 90% OH, CH₃, N, O, N, HCO, O, OAc

In all these reactions, a similar mechanism can be postulated. It involves displacement of a good leaving group (halide, tosylate, or chloroformate) by dimethyl sulfoxide to give a dialkylalkoxysulfonium salt, which can then react with a base to give the aldehyde plus dimethyl sulfide.

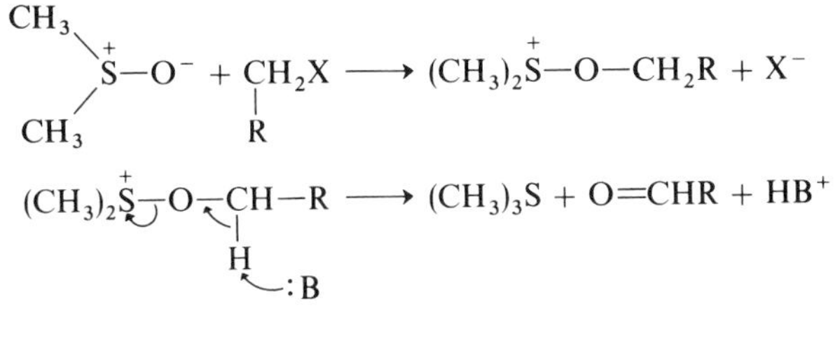

$$(CH_3)_2\overset{+}{S}{-}O^- + \underset{R}{CH_2X} \longrightarrow (CH_3)_2\overset{+}{S}{-}O{-}CH_2R + X^-$$

$$(CH_3)_2\overset{+}{S}{-}O{-}\underset{H\ :B}{CH}{-}R \longrightarrow (CH_3)_3S + O{=}CHR + HB^+$$

A related reaction is that of dimethyl sulfoxide with an epoxide to give an α-hydroxy ketone, from a displacement followed by a base catalyzed elimination.

$$\underset{\diagdown O \diagup}{RCH-CHR} + (CH_3)_2SO \longrightarrow \underset{O^-}{RCH}-\overset{O-\overset{+}{S}(CH_3)_2}{CHR} \longrightarrow \underset{OH}{RCH}-\overset{O}{\overset{\|}{C}}-R + (CH_3)_2S$$

MANGANESE DIOXIDE OXIDATION OF ALLYLIC ALCOHOLS

Most of the methods discussed thus far are applicable to either any primary or any secondary alcohol. We turn now to two reagents that react with only a limited number of alcohols, those specifically activated in one way or another. Manganese dioxide is such a reagent in that it usually reacts only with primary and secondary alcohols adjacent either to carbon-carbon double bonds or to benzene rings. Thus, benzyl alcohol gives benzaldehyde, allyl alcohol gives acrolein, and vitamin A and the alkynenol shown give the corresponding aldehydes.

$$C_6H_5-CH_2OH \xrightarrow{MnO_2} C_6H_5-\overset{O}{\overset{\|}{C}}H$$

$$CH_2=CHCH_2OH \xrightarrow{MnO_2} CH_2=CH\overset{O}{\overset{\|}{C}}H$$

$$HC\equiv CCH=\underset{CH_3}{C}CH_2OH \xrightarrow{MnO_2} HC\equiv CCH=\underset{CH_3}{C}CHO$$

$$\text{vitamin A } (R-CH_2OH) \xrightarrow[\text{pentane, 25°, 7 days}]{MnO_2} R-CHO$$

vitamin A

The mechanism of this reaction is relatively obscure, since manganese dioxide, a solid, is not dissolved in the reagents in which it is employed—usually chloroform, cyclohexane, or other nonhydroxylic solvents—and the precise state of the manganese in the solid is not known. On the other hand, any oxidation normally involves the formation of an intermediate either of carbonium or radical character. One can study this by trapping intermediates or by observing the effect of substituents upon rate. A related reaction of diphenylmethanes is interesting in this respect, in that these compounds are sufficiently activated (by the two phenyl groups) that they are oxidized to diphenyl ketones. Tetraphenylethanes are observed as side reaction products, important because a tetraphenylethane could only be formed by coupling of radical (not carbonium ion) intermediates. Another argument for a radical intermediate is the observation that the yields of ketones are considerably better for *p*-nitrodiphenylmethanes than for *p*-methoxy-; a *p*-nitro group could stabilize a radical intermediate but not a carbonium ion.

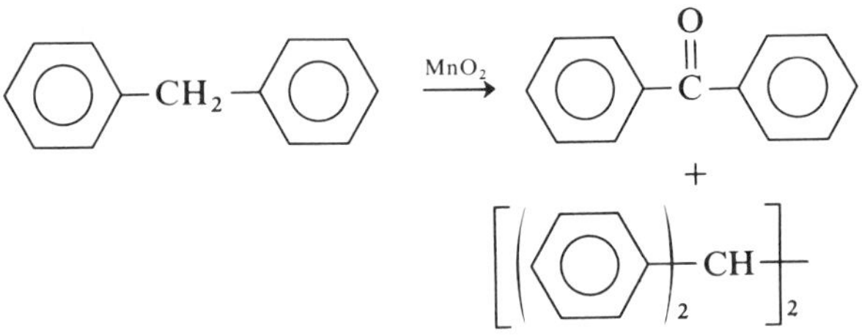

$$C_6H_5-CH_2-C_6H_5 \xrightarrow{MnO_2} C_6H_5-\overset{O}{\overset{\|}{C}}-C_6H_5 + [(C_6H_5)_2CH-]_2$$

BISMUTH OXIDE OXIDATION OF HYDROXY KETONES

Another specific oxidizing agent for activated alcohols is bismuth oxide. This mild reagent converts α-hydroxy ketones to α-diketones, leaving untouched other functional groups such as carbon-carbon double bonds or isolated alcohols.

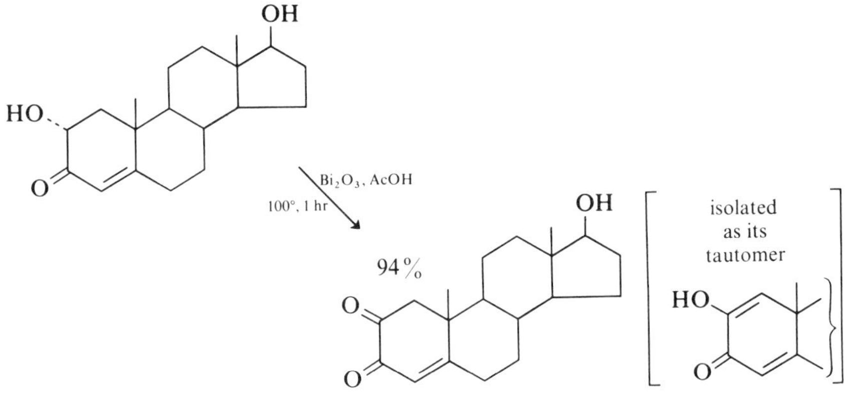

OH

HO

O

Bi_2O_3, AcOH

100°, 1 hr

94%

OH

O

O

isolated as its tautomer

HO

O

SODIUM PERIODATE CLEAVAGE

Sodium periodate is a highly selective reagent that oxidizes vicinal glycol groupings, i.e., 1,2-diols. For obvious reasons, this reagent has been very much used and studied in sugar chemistry, where it serves as one of the favorite quantitative as well as qualitative tools. The reaction can be illustrated by the oxidation of ribitol, 1,2,3,4,5-pentanepentaol. This compound consumes 4 moles of sodium periodate per 1 mole of compound, indicating four pairs of vicinal glycol groups. The reaction is carried out with an excess of periodate, and at the end, the amount consumed is indicated by titration with an excess of arsenite, which reduces periodate to iodate. Excess arsenite, in turn, is titrated to an end point with iodine. The products of the oxidation can be identified both qualitatively and quantitatively and shown to be 3 moles of formic acid and 2 moles of formaldehyde. The structure of ribitol follows directly from this result.

$$\underset{\text{ribitol}}{\begin{array}{c} CH_2OH \\ | \\ H-C-OH \\ | \\ H-C-OH \\ | \\ H-C-OH \\ | \\ CH_2OH \end{array}} + 4IO_4^- \longrightarrow \begin{array}{c} CH_2O \\ + \\ HCOOH \\ + \\ HCOOH \\ + \\ HCOOH \\ + \\ CH_2O \end{array} + 4IO_3^- + H_2O$$

$$IO_4^- + \underset{\text{excess}}{AsO_3^{3-}} \longrightarrow IO_3^- + AsO_4^{3-}$$

$$AsO_3^{3-} + \underset{\text{violet}}{I_2} + H_2O \longrightarrow \underset{\text{colorless}}{AsO_4^{3-} + 2I^- + 2H^+}$$

Other vicinal groupings are also oxidized by periodate, notably α-hydroxy ketones and aldehydes, vicinal amino alcohols, and, more slowly, α-diketones and α-hydroxy acids.

$$CH_3COCOCH_3 + IO_4^- + H_2O \longrightarrow 2CH_3COOH + IO_3^-$$

$$HOCH_2\underset{\substack{| \\ NH_2}}{C}HCOOH + IO_4^- \longrightarrow CH_2O + HCOCOOH + NH_3 + IO_3^-$$

$$CH_3\overset{\substack{CH_3 \\ |}}{\underset{\substack{| \\ OH}}{C}}COCH_3 + IO_4^- \longrightarrow (CH_3)_2CO + HOCOCH_3 + IO_3^-$$

Selectivity of the reagent is demonstrated further by the relative rates of periodate oxidation of cis- and trans-glycols. As illustrated, cis-1,2-cyclohexanediol is oxidized 30 times as fast as the trans isomer, and similar relative rates (k_{rel}) are obtained for cyclopentanediols. These relative rates suggest the intermediacy of a cyclic periodate ester. In further agreement with this, 1,2-diols in which the hydroxyl groups cannot conceivably form a cyclic ester are completely unreactive toward periodate under standard reaction conditions.

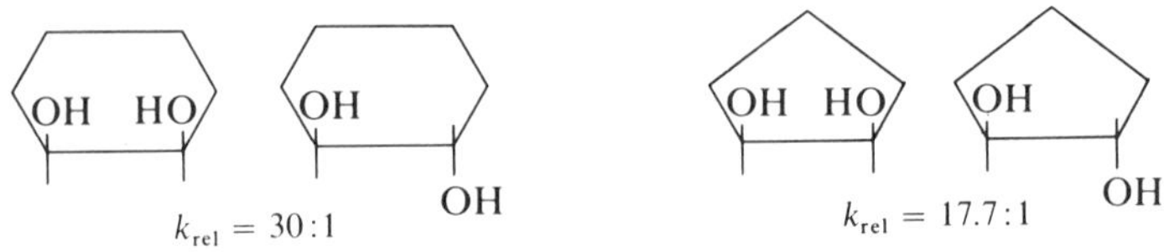

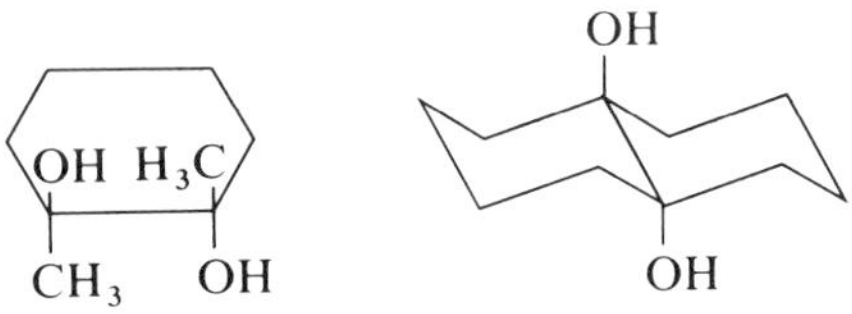

$k_{rel} = 30:1$

$k_{rel} = 17.7:1$

unreactive

unreactive

The mechanism of periodate oxidation is somewhat complicated by the fact that periodate exists in a variety of states of hydration. The mechanism shown is illustrated with the lowest hydration state, IO_4^-, although higher hydration states are probably involved, e.g., $H_2IO_5^-$ or $H_4IO_6^-$. Further indication of a cyclic mechanism is provided by studies with ^{18}O, which demonstrate that the oxygen atoms present in the original hydroxyl groups remain in the carbonyl groups formed. On the other hand, ^{18}O in the periodate is incorporated into acidic products.

$$IO_4^- + 2H_2O \rightleftharpoons H_4IO_6^- \rightleftharpoons H_3IO_6^{2-} \rightleftharpoons H_2IO_6^{3-}$$

$$H_3IO_6^{2-} \rightleftharpoons H_5IO_6$$

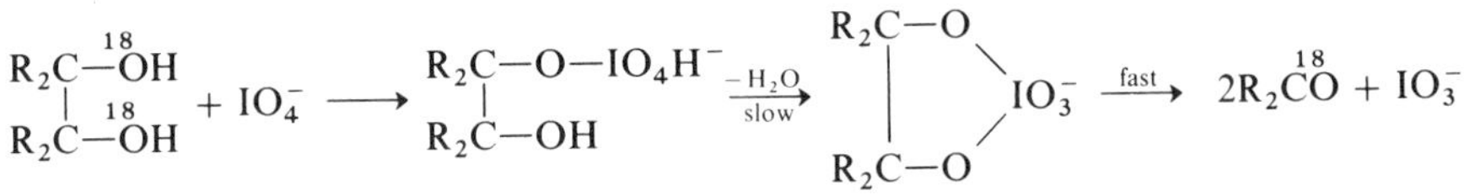

$$R_2C(^{18}OH)\text{—}C(^{18}OH)R_2 + IO_4^- \longrightarrow R_2C(\text{—O—}IO_4H^-)\text{—}C(OH)R_2 \xrightarrow[\text{slow}]{-H_2O} \text{cyclic } (R_2C\text{—O})_2IO_3^- \xrightarrow{\text{fast}} 2R_2C{=}^{18}O + IO_3^-$$

$$CH_3\overset{O}{\overset{\|}{C}}\text{—}\overset{O}{\overset{\|}{C}}CH_3 + \underset{(^{18}O)}{IO_4^-} + H_2O \longrightarrow 2CH_3C^{18}OOH + IO_3^-$$

PERIODATE-PERMANGANATE AND PERIODATE-OSMIUM TETROXIDE REAGENTS

Since alkenes are converted to glycols by permanganate or osmium tetroxide, and glycols are cleaved by periodate, combinations of permanganate or osmium tetroxide with periodate provide pairs of very effective oxidizing agents for alkenes. Procedures employing these reagents compete effectively with the ozonolysis procedure for alkene cleavage, since no special equipment (ozonizer) is required. Collectively, the reagents are known as the Lemieux-von Rudloff reagents. As expected, either permanganate or osmium tetroxide oxidizes the alkene to a glycol, and the glycol reacts selectively with periodate to give two new carbonyl groups. Aldehydes and ketones are obtained from the osmium tetroxide-periodate combination. Potassium permanganate, however, is a sufficiently vigorous reagent to oxidize the aldehydes further to acids, and these, plus ketones, are the products with the permanganate-periodate combination. An interesting aspect of these combination reagents is that sodium periodate is a strong enough oxidizing agent to reconvert the reduced osmium product to osmium tetroxide and the lower oxidation states of manganese to permanganate.

$$CH_3CH_2CH{=}\underset{\displaystyle CH_3}{\underset{|}{C}}CH_2CH_3 \xrightarrow[\substack{KMnO_4\\K_2CO_3\\H_2O}]{NaIO_4} \xrightarrow{H^+} CH_3CH_2COOH + CH_3COCH_2CH_3$$

$$\xrightarrow[NaIO_4]{OsO_4} CH_3CH_2CHO + CH_3COCH_2CH_3$$

CLEAVAGE BY LEAD TETRAACETATE

An alternative oxidizing agent for 1,2-diols is lead tetraacetate. The principal choice between lead tetraacetate and sodium periodate as a reagent for effecting α-glycol cleavage rests on whether the reaction is to be carried out in water, in which periodate is more effective, or in organic solvents, in which lead tetraacetate dissolves. Lead tetraacetate, however, is less selective and oxidizes a number of other functional groups, converting, for example, alkenes to α-acetoxyalkenes, as we saw in Chapter 3. The general mechanism of oxidation of vicinal glycols by lead tetraacetate,

$$\begin{array}{c} CO_2C_4H_9 \\ | \\ H-C-OH \\ | \\ H-C-OH \\ | \\ CO_2C_4H_9 \end{array} \xrightarrow[\substack{C_6H_6,\ 30^\circ\\1\ hr}]{Pb(OAc)_4} 2H-\overset{\displaystyle CO_2C_4H_9}{\overset{|}{C}}{=}O + Pb(OAc)_2 \quad 82\%$$

$$R\underset{\displaystyle OH}{\underset{|}{C}}HCOOH \xrightarrow{Pb(OAc)_4} RCHO + CO_2$$

$$RCO-COR \longrightarrow 2RCOOH$$

$$R\underset{\displaystyle OH}{\underset{|}{C}}HCOR \longrightarrow RCHO + HOCOR$$

like that by periodate, probably involves a cyclic intermediate, which is indicated by the relative rates of oxidations of most cis- and trans-1,2-diols.

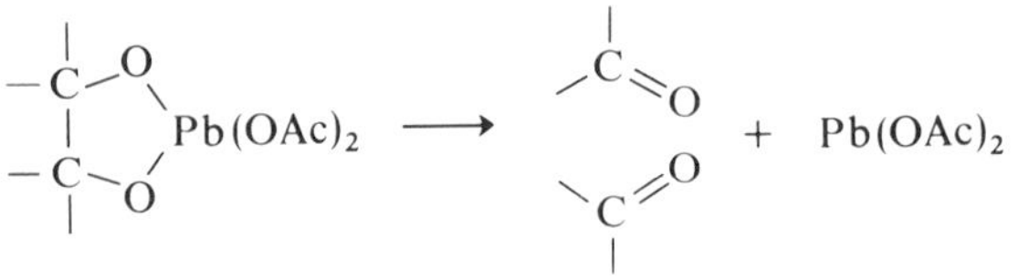

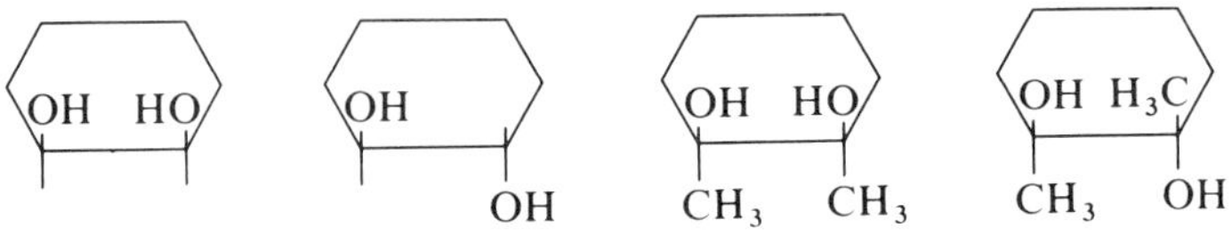

$Pb(OAc)_2 \longrightarrow$ C=O, C=O $+ \; Pb(OAc)_2$

OH HO OH OH

$k_{rel} = 20:1$

OH HO, CH_3 CH_3; OH H_3C, CH_3 OH

$k_{rel} = 20:1$

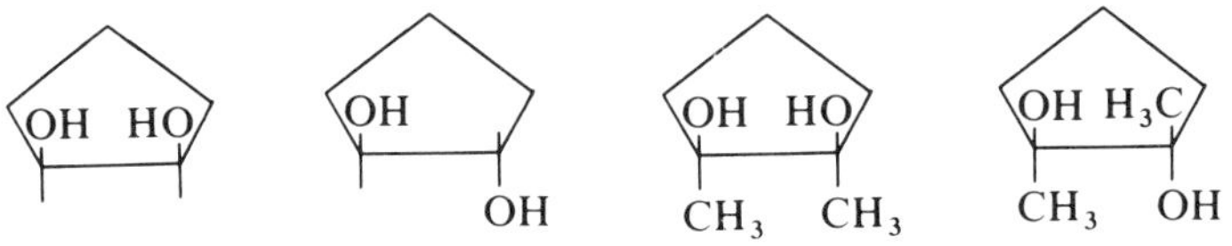

OH HO OH OH

$k_{rel} = 3000:1$

OH HO, CH_3 CH_3; OH H_3C, CH_3 OH

$k_{rel} = 200:1$

OH HO OH OH

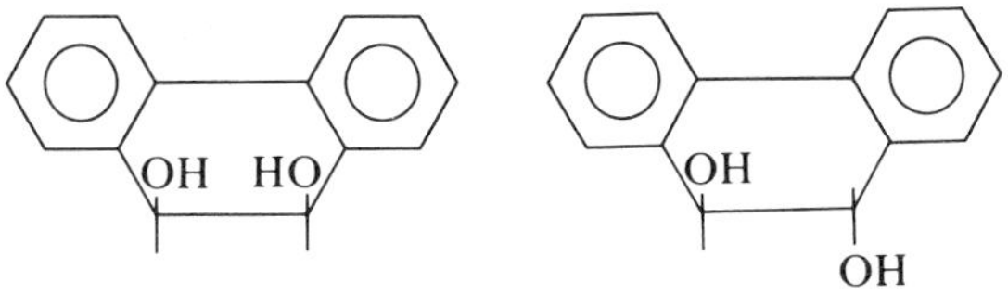

$k_{rel} = 10:1$

An alternative to the cyclic mechanism must exist, however, since some trans-1,2-diols that do not react at all with periodate react, albeit slowly, with tetraacetate. The alternative mechanism shown involves attack of base on an acyclic lead derivative.

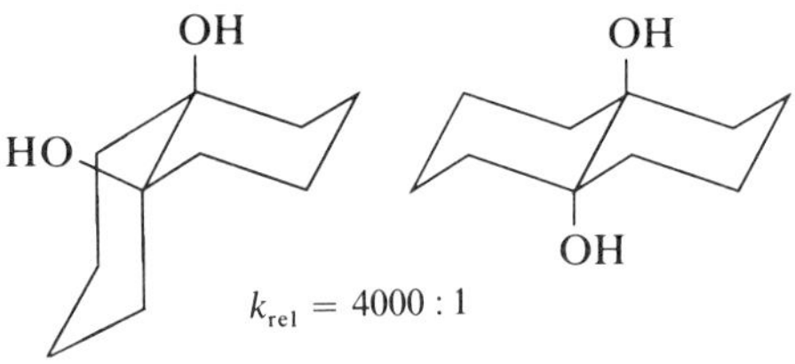

OH HO OH OH

$k_{rel} = 4000:1$

$$\text{B:}\;\; \text{HO}-\text{C}-\text{C}-\text{OPb(OAc)}_3 \longrightarrow \text{B}-\text{H}^+ + \text{O}=\text{C} + \text{C}=\text{O} + \text{Pb(OAc)}_2 + {}^-\text{OAc}$$

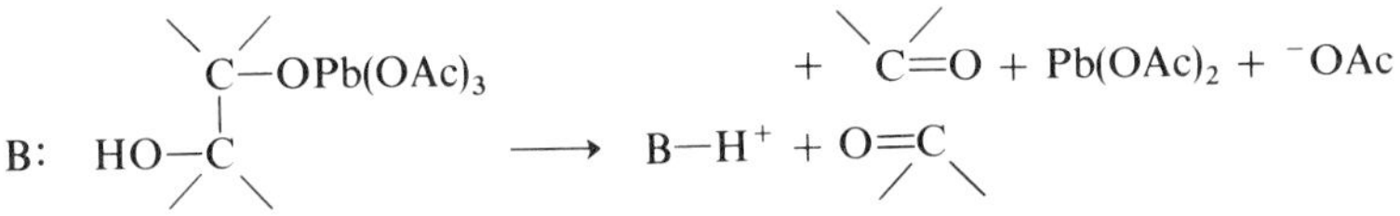

Studies carried out with substituted benzpinacols, having hydroxyl groups on benzylic carbon atoms, indicate electron demand in the transition state, since electron-donating substituents speed the reaction.

$$\left[\left(X\text{-}C_6H_4\right)_2\overset{OH}{\underset{|}{C}}-\right]_2$$

A reaction related to glycol cleavage is the decarboxylation of 1,2-dicarboxylic acids by lead tetraacetate. Again, a cyclic intermediate can be written, although the evidence for its formation is less compelling. The product is an alkene, and a decision as to whether oxidation or reduction of the organic reactant, 1,2-cyclohexanedicarboxylic acid, is taking place in the reaction can be made by observing that the inorganic reagent, lead tetraacetate, has been reduced to lead diacetate; therefore, the organic reactant must have been oxidized.

$$\text{1,2-}C_6H_{10}(COOH)_2 \xrightarrow{Pb(OAc)_4} \text{cyclohexene} + 2CO_2 + Pb(OAc)_2 + 2HOAc$$

AUTOXIDATION OF ETHERS

Most of the reactions we have discussed for oxidizing alcohols to aldehydes or ketones have been two-electron oxidations, and many of them have involved the formation of intermediate inorganic esters, such as that shown on page 71. The requirement of an ester intermediate eliminates the possibility of oxidation of ethers, since they lack the hydroxyl group required to form esters. Thus, ethers are relatively stable to most oxidizing agents. On the other hand, ethers should be susceptible to oxidation by an oxidizing agent that reacts at the carbinol hydrogen, the $-\underset{|}{C}HO-$ hydrogen.

Such a reagent is oxygen, and ethers are notorious for the formation of peroxides in the presence of air, a process called autoxidation. These peroxides are almost never isolated, nor are they useful, but they build up and have been the cause of violent, sometimes fatal, explosions that usually occur when the ether is being distilled. Since the peroxides are higher boiling, they concentrate in the distilling flask until their detonation temperature is reached. Especially dangerous is the peroxide of diisopropyl ether, but even diethyl ether forms an explosive compound. The structures of these peroxides are believed to be as shown. When they decompose, as in all initiation steps, they form two radicals which can trigger additional decompositions, giving a rapid buildup of volatile products (i.e., an explosion). One should always be suspicious of ether that has stood for a prolonged period on the shelf. Peroxides can be removed from this solvent either by reduction with lithium aluminum hydride or by washing with ferrous sulfate solution. Ethers are often stored with iron wire present as an additional precaution.

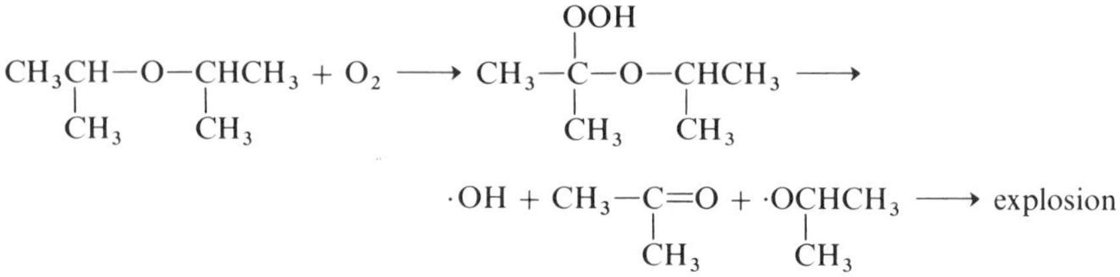

$$CH_3\underset{CH_3}{CH}-O-\underset{CH_3}{CHCH_3} + O_2 \longrightarrow CH_3-\overset{OOH}{\underset{CH_3}{C}}-O-\underset{CH_3}{CHCH_3} \longrightarrow$$

$$\cdot OH + CH_3-\underset{CH_3}{C}=O + \cdot O\underset{CH_3}{CHCH_3} \longrightarrow \text{explosion}$$

AUTOXIDATION OF PHENOLS

Phenols, whose hydroxyl group is modified by the attached benzene ring, behave somewhat differently toward oxidizing agents than alcohols. They are much easier to oxidize than alcohols and are especially susceptible to autoxidation, which leads initially to a resonance-stabilized phenoxy radical. Like other radicals, phenoxy radicals attack benzene rings, in effect giving coupling products, which can be further oxidized. Radical oxidation of phenol itself is especially complex, but a number of products have been isolated from ortho- and para-substituted phenols. For example, *p*-cresol gives a simple ortho-coupling product as well as the more complex Pummerer's ketone, and 2,6-dimethylphenol gives a para-coupling product, which is then further oxidized to the compound shown. Since the latter compound has a highly extended chromophoric system of π-electrons, it is, as anticipated, a colored compound. Similar compounds apparently form with unsubstituted phenols but are more difficult to identify. A bottle of a phenol that has been exposed to air is very likely to be colored, due to the formation of these complex colored products.

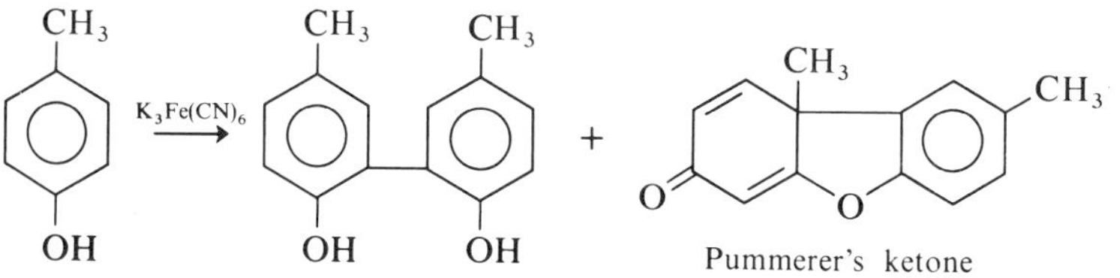

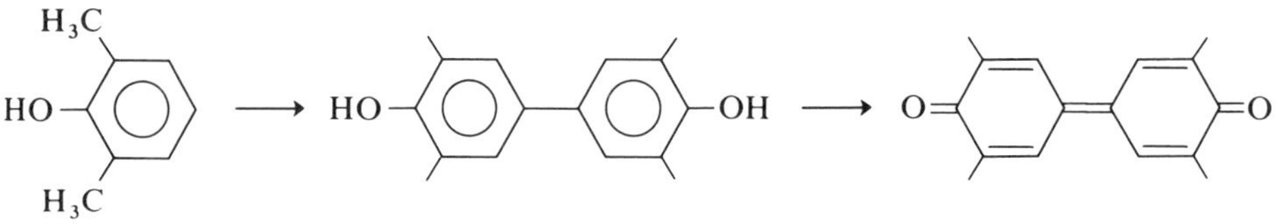

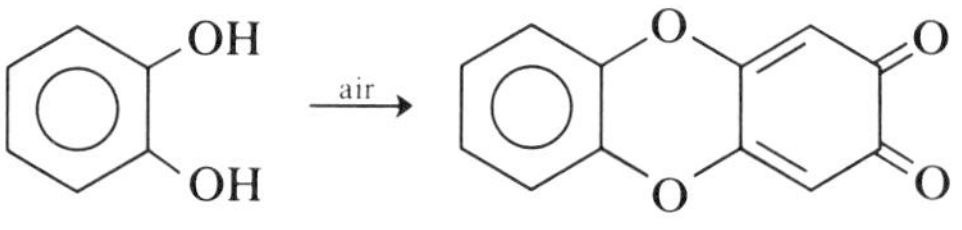

Similar one-electron oxidations of phenols can also be carried out under more controlled conditions by other reagents, such as potassium ferricyanide or alkyl peroxides.

HO–C₆H₄(CH₃) $\xrightarrow{K_3Fe(CN)_6}$ O=(CH₃)C₆H₃=C₆H₃(CH₃)=O

$$\text{2,6-}[(CH_3)_3C]_2\text{-4-}CH_3C_6H_2OH \xrightarrow[OH^-]{K_3Fe(CN)_6} O{=}C_6H_2[C(CH_3)_3]_2{=}CH{-}CH{=}C_6H_2[C(CH_3)_3]_2{=}O$$

$$\xrightarrow{[(CH_3)_3CO]_2} O{=}C_6H_2[C(CH_3)_3]_2(CH_3)OOC(CH_3)_3 \longrightarrow O{=}C_6H_2[C(CH_3)_3]_2{=}O$$

The ease of oxidation of phenols allows their use where antioxidants are needed as scavengers of radicals to prevent chain reactions. Particularly effective in this respect is 2,6-di-*t*-butyl-4-methylphenol, which is a commercial antioxidant used to prevent polymerization of alkenes as well as oxidation of highly susceptible compounds. This phenol, with neither free ortho nor free para positions, cannot couple like the phenols already noted and is relatively stable. When it does react, it often does so by loss of alkyl groups, as shown in the third equation above.

OXIDATION OF PHENOLS TO QUINONES

Although a complex mixture of oxidation products is often obtained from a phenol, it is often possible to convert a phenol to a single product, usually the corresponding quinone. Complicated binuclear quinones were products of autoxidation and controlled oxidation on page 86 and above, but phenols also can form simple quinones, often in reasonable yield. For example, 2,3,6-trimethylphenol is oxidized by sodium dichromate to 2,3,5-trimethylbenzoquinone.

$$\text{2,3,6-}(CH_3)_3C_6H_2OH \xrightarrow[H_2SO_4]{Na_2Cr_2O_7} 50\% \text{ 2,3,5-trimethylbenzoquinone}$$

An especially useful reagent is Fremy's salt, which attacks the phenol, giving initial formation of an oxide radical. The radical then reacts with a second mole of Fremy's salt to give an intermediate, which with base gives the quinone in reasonable yield.

$\cdot O-N(SO_3K)_2$ + C_6H_5–OH ⟶ C_6H_5–O·

Fremy's salt

$\cdot ON(SO_3K)_2$

O=C_6H_4(H)–O–$N(SO_3K)_2$

O=C_6H_4(H)–O–$N(SO_3K)_2$ + ^{-}OH ⟶ O=C_6H_4=O

It is much easier to oxidize 1,2- and 1,4-dihydroxybenzenes or their corresponding amine analogues to quinones, as can be illustrated in the following two equations.

$NH_3\cdot HCl$, OH (naphthalene) $\xrightarrow{FeCl_3}$ 94% 1,2-naphthoquinone (O, O)

H_3C, OH, OH (benzene) $\xrightarrow{Ag_2O}$ H_3C (quinone, O, O)

Actually, the oxidation of a 1,2- or 1,4-dihydroxybenzene to a quinone and its reverse, the reduction of a quinone, is reversible, one of very few reactions found in organic chemistry that are reversible under mild conditions. For this reason, quinones play an important role in the oxidation and reduction of biological compounds. The quinone group is also important as a chromophore, since many quinones, e.g., hydroxyanthraquinones, are colored and are found in nature or have been synthesized; their reduction products, the hydroquinones, are colorless and are called leuco forms.

Oxidation and reduction of quinones and hydroquinones are sufficiently reversible that they can be carried out at an electrode, and a quantitative measure of the ease of reduction of the quinone is obtained by the redox potential expressed in volts. A number of quinones are compared in this way in Table 5-1. In this table, those quinones, e.g. *o*-benzoquinone, which are easiest to reduce to the hydroquinones have the highest redox potentials. By contrast, those, e.g. anthraquinone, which are most stable have the lowest redox potentials.

Table 5-1

REDOX POTENTIALS OF QUINONES

Quinone	Redox potential, volts (alcohol)
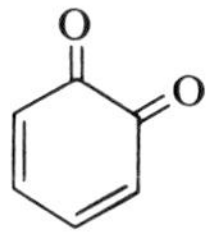	0.794
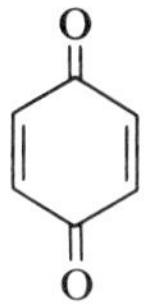	0.715
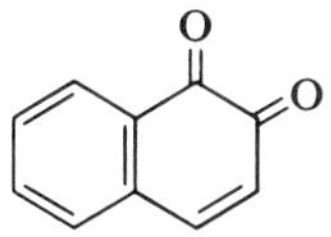	0.576
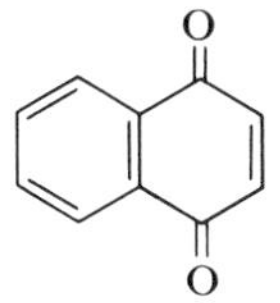	0.484
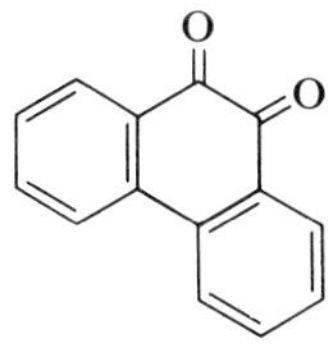	0.460
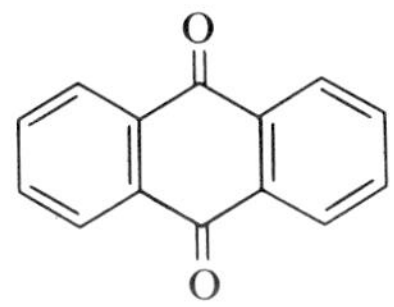	0.154

Table 5-2

RELATIVE REDOX POTENTIALS OF 2-SUBSTITUTED NAPHTHOQUINONES

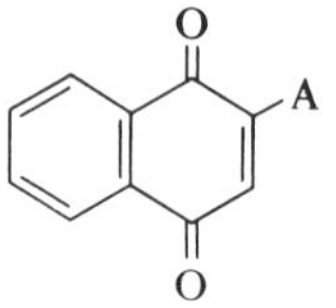

A (2-substituent of 1,4-naphthoquinone)	ΔV, volts (relative to naphthoquinone)
A = $-SO_2-C_6H_4-CH_3$ 	+0.121
$-SO_3Na$	+0.069
$-Cl$	+0.024
$-OAc$	−0.009
 $-C_6H_5$	−0.032
$-CH_3$	−0.076
$-OH$	−0.128
$-NH_2$	−0.252

We see in Table 5-2 that the redox potential of quinones is susceptible to modification by the introduction of substituents. In general, electron-donating substituents lower the redox potential, electron-withdrawing substituents raise it. If one desires a highly oxidizing quinone, one substitutes electron-withdrawing groups. From this, it is no surprise that the quinones which we saw to be most reactive in dehydrogenating cyclic hydrocarbons in Chapter 2 are those which are highly substituted with chloro, fluoro, or cyano groups.

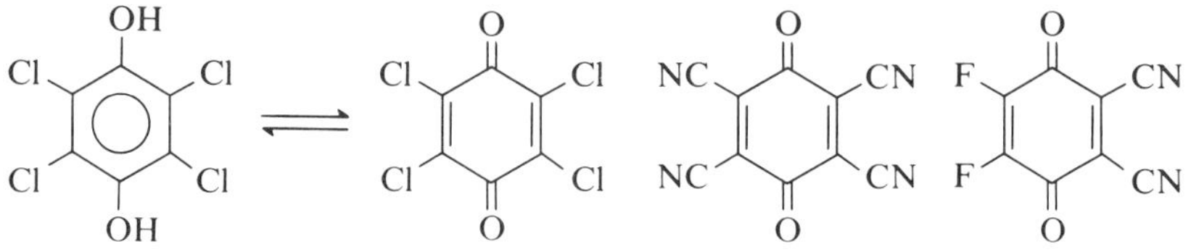

REFERENCES

Augustine (*Oxidation*): pp. 189–212, A. S. Perlin, "Glycol Cleavage and Related Oxidations."

Cullis, C. F., and A. Fish, "Carbonyl-Forming Oxidations," in *Chemistry of the Carbonyl Group*, S. Patai, ed., John Wiley & Sons, Inc., New York, 1966, Chap. 2.

Djerassi, C., "The Oppenauer Oxidation," *Org. Reactions*, **6**, 207 (1951).

Epstein, W. W., and F. W. Sweat, "Dimethyl Sulfoxide Oxidations," *Chem. Rev.*, **67**, 247 (1967).

Evans, R. M., "Oxidation by Manganese Dioxide in Neutral Media," *Quart. Rev.* (London), **13**, 61 (1959).

Foerst (Vol. 2): pp. 303–36, K. Heyns and H. Paulsen, "Selective Catalytic Oxidations with Noble Metal Catalysts."

Heyns, K., H. Paulsen, G. Rüdiger, and J. Weyer, "Configurational and Conformational Selectivity in the Catalytic Oxidation with Oxygen on Platinum Catalysts," *Fortschr. Chem. Forsch.*, **11**, 285 (1969).

Musso, H., "Oxidation Reactions of Phenols," *Angew. Chem.*, **75**, 965 (1963).

Stewart: Chap. 8, "Nonmetal Oxides and Acids."

Taylor, W. I., and A. R. Battersby, eds., *Oxidative Coupling of Phenols*, Marcel Dekker, Inc., New York, 1967.

Waters: Chap. 4, "The Oxidation of Alcohols."

Waters: Chap. 9, "Oxidations of Phenols and Aromatic Amines."

Wiberg: Chap. 2, C. A. Bunton, "Glycol Cleavage and Related Reactions."

Wiberg: Chap. 5, R. Criegee, "Oxidations with Lead Tetraacetate."

Wiberg: Chap. 1, R. Stewart, "Oxidation by Permanganate."

Wiberg: Chap. 2, K. B. Wiberg, "Oxidation by Chromic Acid and Chromyl Compounds."

PROBLEMS

1. Write a detailed mechanism for each of the following reactions, including all intermediates and/or transition states.

(a) $$CH_3CH(OH)CH_3 \xrightarrow[H_2SO_4]{CrO_3} CH_3COCH_3$$

(b) cyclohexane-1,2-dicarboxylic acid (COOH, COOH) $\xrightarrow{Pb(OAc)_4}$ cyclohexene

(c) $$CH_3CH(OH)CH(OH)CH_2CH_2CH_3 \xrightarrow{NaIO_4} CH_3{-}\overset{O}{\overset{\|}{C}}H + H\overset{O}{\overset{\|}{C}}{-}CH_2CH_2CH_3$$

(d) 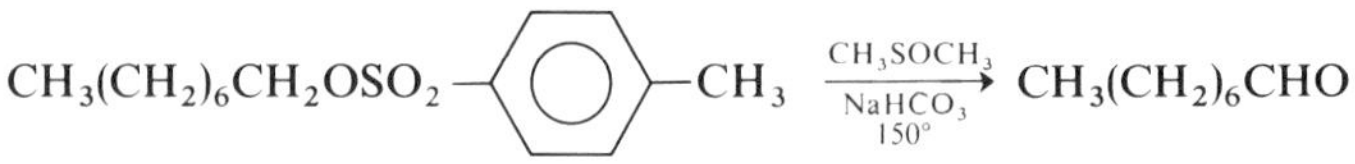 $CH_3(CH_2)_6CH_2OSO_2-C_6H_4-CH_3 \xrightarrow[NaHCO_3,\ 150°]{CH_3SOCH_3} CH_3(CH_2)_6CHO$

2. Suggest reaction conditions (reagents, catalysts, solvents, etc.) suitable for effecting the following conversions. More than one step may be required.

(a) 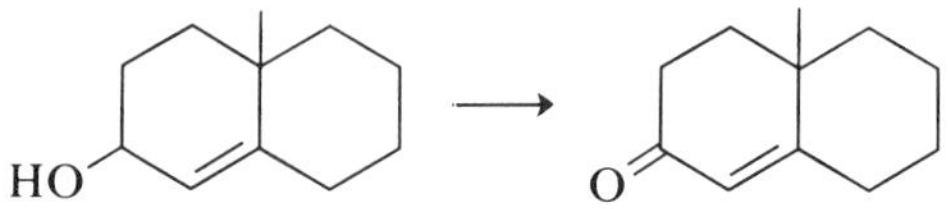 HO– (bicyclic alcohol) ⟶ O= (bicyclic enone)

(b) 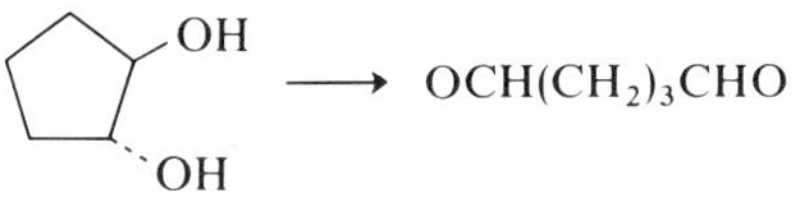 cyclopentane-1,2-diol (OH, OH) ⟶ $OCH(CH_2)_3CHO$

(c) 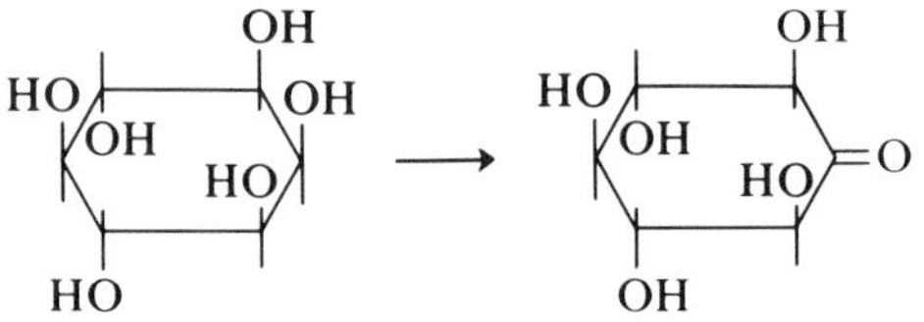 HO, OH, OH, OH, HO, HO (cyclohexanehexol) ⟶ HO, OH, OH, HO, =O, OH

(d) 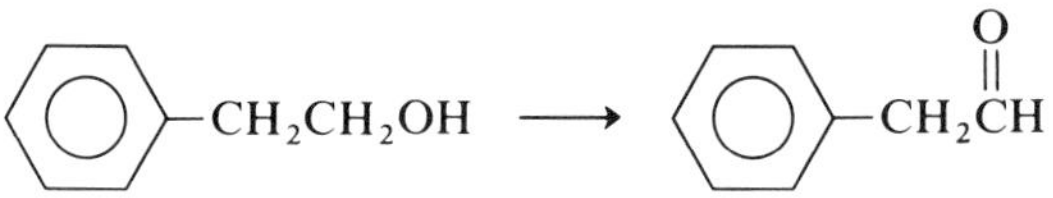 $C_6H_5-CH_2CH_2OH \longrightarrow C_6H_5-CH_2\overset{O}{\overset{\|}{C}}H$

(e) 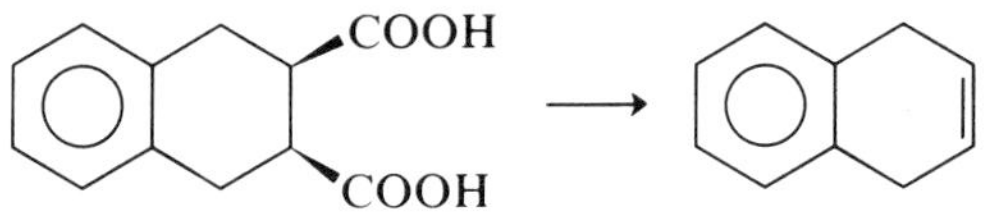 COOH, COOH ⟶

(f) 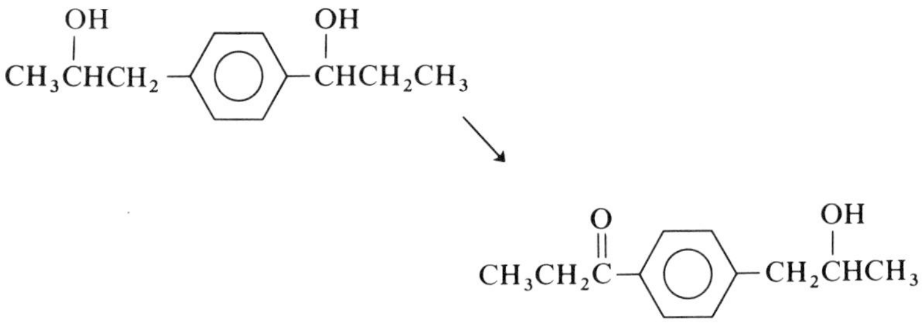 $CH_3\underset{}{\overset{OH}{\overset{|}{C}}}HCH_2-C_6H_4-\overset{OH}{\overset{|}{C}}HCH_2CH_3$

⟶ $CH_3CH_2\overset{O}{\overset{\|}{C}}-C_6H_4-CH_2\overset{OH}{\overset{|}{C}}HCH_3$

3. Indicate all the expected products of the following reactions.

(a) 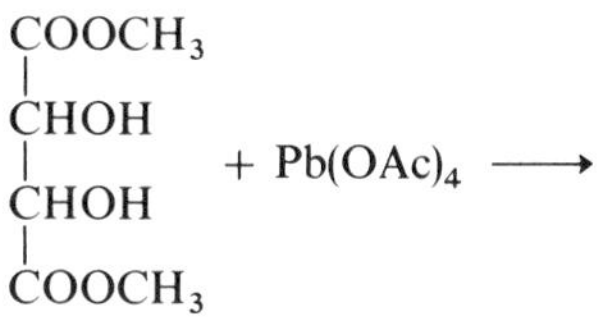 $COOCH_3$–CHOH–CHOH–$COOCH_3$ + $Pb(OAc)_4$ ⟶

(b) 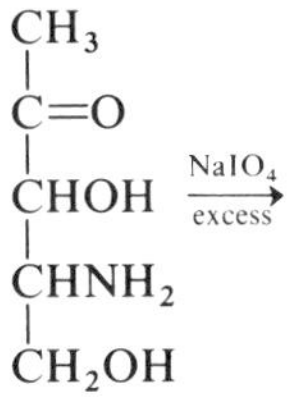 CH_3–C=O–CHOH–$CHNH_2$–CH_2OH $\xrightarrow[\text{excess}]{NaIO_4}$

(c) (1-hydroxy-1,2,3,4-tetrahydronaphthalene: H, OH; H_2, H_2, H_2) + MnO_2 $\xrightarrow{\Delta}$

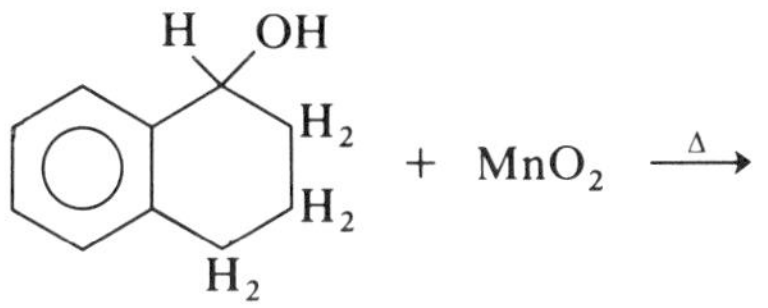

4. Write balanced equations for the oxidation steps of Problems 1(b), 1(c), 1(d), 2(a), and 2(c).

5. Show all the intermediates in the oxidations of oleic acid by sodium periodate-potassium permanganate and by osmium tetroxide-sodium periodate.

6. Explain by means of appropriate equations or formulas why 2,6-di-*t*-butyl-4-methylphenol is sometimes added as an antioxidant to gasoline.

7. Which of the following compounds will be oxidized most rapidly by chromic oxide? Why?

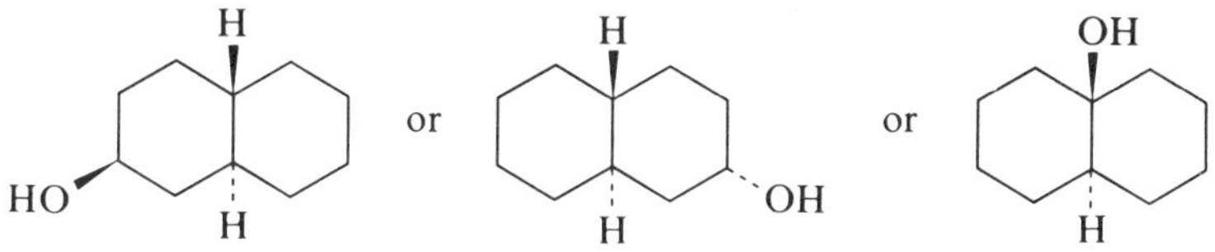

(decalin with H, HO, H) or (decalin with H, OH, H) or (decalin with OH, H)

6

Reduction of Carbonyl Compounds to Alcohols

LITHIUM ALUMINUM HYDRIDE REDUCTION

The laboratory reagent most widely used today for the conversion of carbonyl compounds to alcohols is lithium aluminum hydride. It is a very powerful reagent, one which can reduce even a carboxyl group to a primary alcohol. A summary of the types of carbonyl groups that are reduced by

Table 6-1

TYPES OF COMPOUNDS REDUCED BY LITHIUM ALUMINUM HYDRIDE

Reactant	Example	Product
Aldehyde	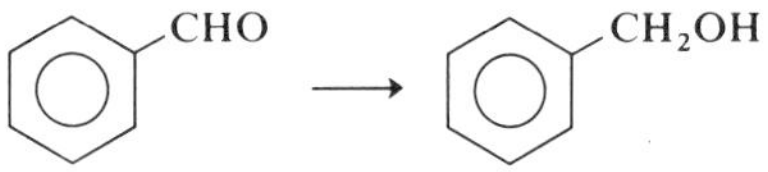 $C_6H_5CHO \longrightarrow C_6H_5CH_2OH$	1° alcohol
Ketone	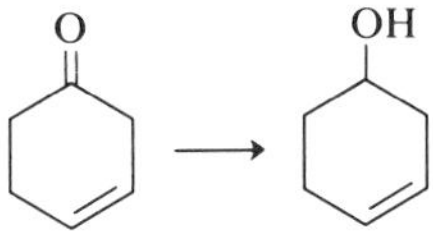 cyclohex-3-enone (C=O) $\longrightarrow$ cyclohex-3-enol (CH–OH)	2° alcohol
Acid	$CH_3CH_2COOH \longrightarrow CH_3CH_2CH_2OH$	1° alcohol
Ester	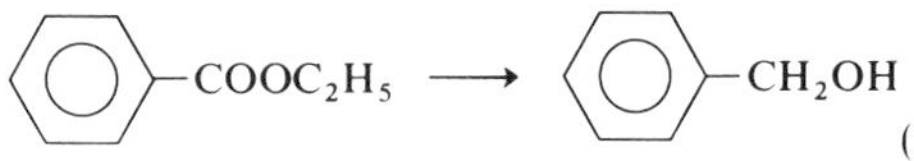 $C_6H_5{-}COOC_2H_5 \longrightarrow C_6H_5{-}CH_2OH$ $(+C_2H_5OH)$	1° alcohol (+ a second alcohol)
Acid chloride	$CH_3{-}C(CH_3)_2{-}COCl \longrightarrow CH_3{-}C(CH_3)_2{-}CH_2OH$	1° alcohol
Anhydride	$(CH_3CO)_2O \longrightarrow CH_3CH_2OH$	1° alcohol

lithium aluminum hydride and the products obtained is found in Table 6-1. Lithium aluminum hydride reductions of nitrogen-containing functional groups are discussed in Chapters 7 and 8.

The reaction usually is carried out at room temperature or below in an ether solvent (not in an alcohol or water), but it can also be carried out in refluxing ethers—diethyl ether or tetrahydrofuran—for difficultly reducible compounds. Although the reagent is expensive, it gives 4 moles of hydride ion per 1 mole of reagent and is, therefore, on a molar basis not exorbitant. It does, however, react with alcohols as well as with other so-called active hydrogen groups (OH, NH, SH), and allowance must be made for those reactions in calculating the theoretical amount required. Usually a convenient excess of reagent is employed.

CHO, HO, CH_3, CH_3 $\xrightarrow{LiAlH_4}$ 29% CH_2OH, HO, CH_3, CH_3

$$CCl_3COCCl_3 \xrightarrow{LiAlH_4} 97\%\ CCl_3CH(OH)CCl_3$$

C_6H_5–CH(N)–CH(N)–$COCH_3$ $\xrightarrow{LiAlH_4}$ 99% C_6H_5–CH(N)–CH(N)–CH(OH)CH_3 (N = morpholino, O in ring)

(furan, O) COOH $\xrightarrow{LiAlH_4}$ 85% (furan, O) CH_2OH

Following a reaction, any excess lithium aluminum hydride must be decomposed carefully, because the reagent reacts violently with water. For this purpose, dropwise addition of ethyl acetate, acetic acid, ethyl alcohol, or isopropyl alcohol is often employed. The reagent is selective enough that it usually does not react with groups other than carbonyl groups in the compound being reduced. However, a conjugated carbonyl compound is sometimes converted to the saturated alcohol instead of to the allylic alcohol, and a few other groups such as nitro groups (see Chapter 8) are also reduced.

$$C_6H_5\text{–}CH{=}CHCOOH \xrightarrow{LiAlH_4} C_6H_5\text{–}CH_2CH_2CH_2OH$$

$$C_6H_5{-}C{\equiv}CCHO \xrightarrow{LiAlH_4} 57\%\ C_6H_5{-}CH{=}CHCH_2OH$$

$$C_6H_{11}{-}C{\equiv}CCOOH \xrightarrow{LiAlH_4} 81\%\ C_6H_{11}(H)C{=}C(H)CH_2OH$$

$$C_6H_5{-}CH{=}CHCHO \xrightarrow[\text{direct addition}]{LiAlH_4,\ r.t.} C_6H_5{-}CH_2CH_2CH_2OH$$

These usually undesirable side reactions can often be eliminated by conducting the reaction at a lower temperature, by employing a limited amount of lithium aluminum hydride, or by adding the hydride solution dropwise to a solution of the carbonyl compound (so-called inverse addition).

$$C_6H_5{-}CH{=}CHCHO \xrightarrow[\text{inverse addition}]{LiAlH_4,\ -10^\circ} C_6H_5{-}CH{=}CHCH_2OH$$

$$C_6H_5(H)C{=}C(CH_3)COOH \xrightarrow[5^\circ]{LiAlH_4} 83\%\ C_6H_5(H)C{=}C(CH_3)CH_2OH$$

$$CH_2{=}CHCOOH \xrightarrow{LiAlH_4} 68\%\ CH_2{=}CHCH_2OH$$

(retinal, with H_3C, CH_3, CH_3, CH_3, CH_3 substituents and terminal $\overset{O}{\|}CH$) $\xrightarrow[r.t.]{LiAlH_4}$ 70% (retinol, terminal CH_2OH)

$$O_2N{-}C_6H_4{-}COOC_2H_5 \xrightarrow[\text{calcd amt inverse addition}]{1/2\ LiAlH_4} O_2N{-}C_6H_4{-}CH_2OH$$

Lithium aluminum hydride can be written $Li^+AlH_4^-$, and the mechanism of lithium aluminum hydride reduction can be envisioned as involving a transfer of hydride ion to the more positive end of a carbonyl group and AlH_3 to the more negative end, as shown.

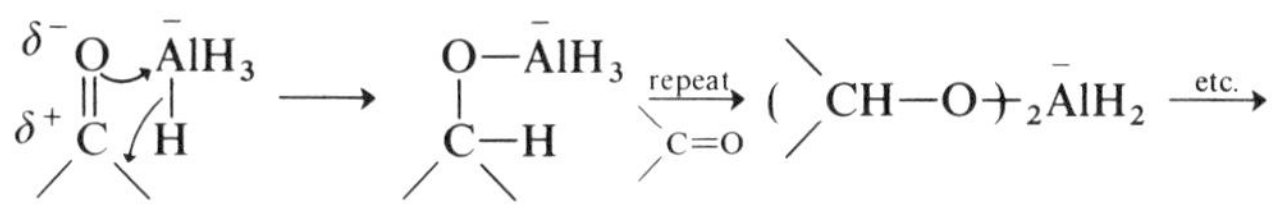

δ^- O—$\bar{A}lH_3$, δ^+ C—H → O—$\bar{A}lH_3$ / C—H $\xrightarrow[\text{C=O}]{\text{repeat}}$ (CH—O$)_2\bar{A}lH_2$ $\xrightarrow{\text{etc.}}$

Since the reagent is reasonably bulky, a sizable steric effect can be observed. In cases such as a rigid 3-keto steroid system, the highly favored product is the 3β-hydroxy (equatorial) compound. This product is favored both by its own stereochemistry, since it is the more stable isomer, and also by the ease of approach of the lithium aluminum hydride reagent from the less hindered bottom side of the steroid.

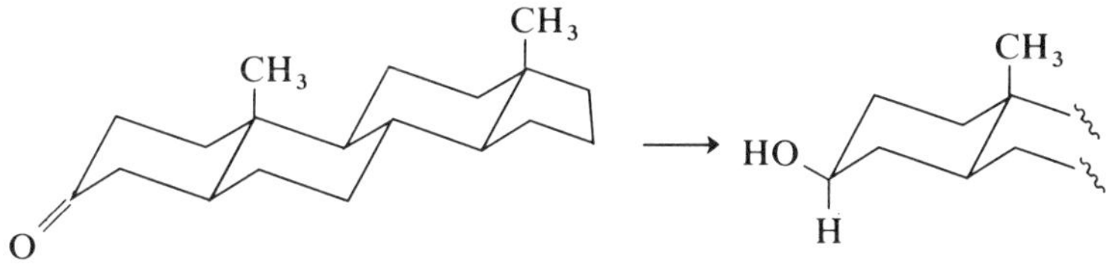

CH_3 CH_3 O → HO H CH_3

In some cases, however, the two effects—relative stability of the product (manifest in equilibrium control) and easier approach of the reagent (kinetic control)—compete, as seen in the reductions of camphor and of 4-*t*-butylcyclohexanone. With camphor the kinetic effect prevails, while with 4-*t*-butylcyclohexanone the equilibrium effect is predominant.

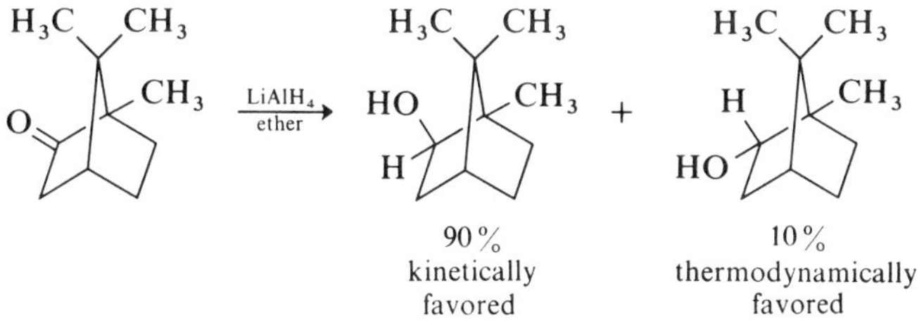

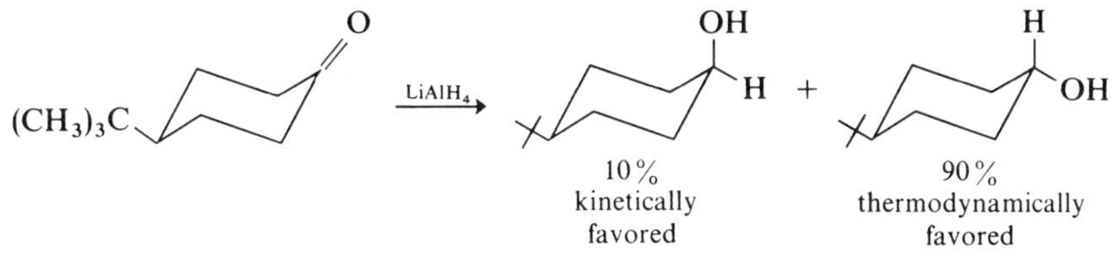

H_3C CH_3 CH_3 O $\xrightarrow[\text{ether}]{\text{LiAlH}_4}$ H_3C CH_3 HO CH_3 H + H_3C CH_3 H CH_3 HO

90%
kinetically
favored

10%
thermodynamically
favored

$(CH_3)_3C$ O $\xrightarrow{\text{LiAlH}_4}$ OH H + H OH

10%
kinetically
favored

90%
thermodynamically
favored

The equilibrium effect can often be enhanced further by employing a bulkier reagent, lithium aluminum hydride plus aluminum chloride, with excess ketone. The excess ketone serves to establish that the equilibrium has been achieved between the two isomeric alcohols, whereas the product complex, bulkier than the alcohol itself, enhances the yield of the equatorial isomer. Excess reagent gives more of the kinetically favored product.

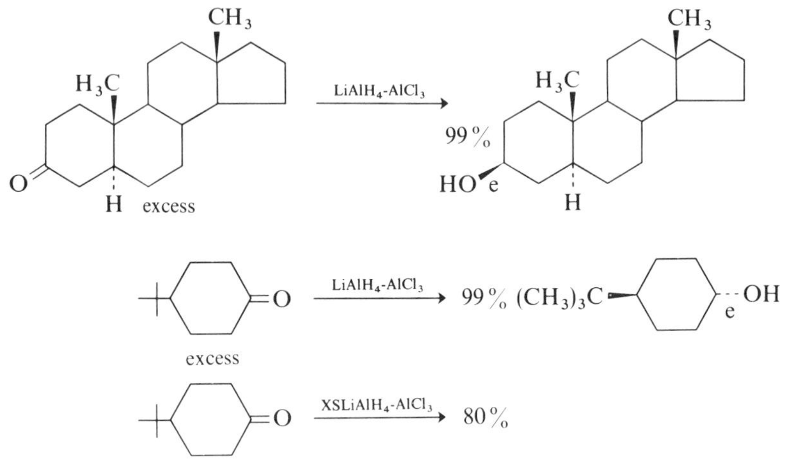

$\xrightarrow{LiAlH_4\text{-}AlCl_3}$ 99% (excess)

$\xrightarrow{LiAlH_4\text{-}AlCl_3}$ 99% $(CH_3)_3C$–(cyclohexyl)–OH (e) (excess)

$\xrightarrow{XSLiAlH_4\text{-}AlCl_3}$ 80%

With acyclic systems, the stereoselectivity of the reduction is somewhat less, but usually the favored isomer is that which is obtained by allowing the ketone to react in the conformation shown (bulky carbonyl group between the two least bulky substituents on the adjacent carbon), with lithium aluminum hydride approaching the carbonyl group from the less hindered side. Nevertheless, both isomers are formed, and one can, in general, expect to obtain both isomers from any reduction with lithium aluminum hydride.

Acyclic:

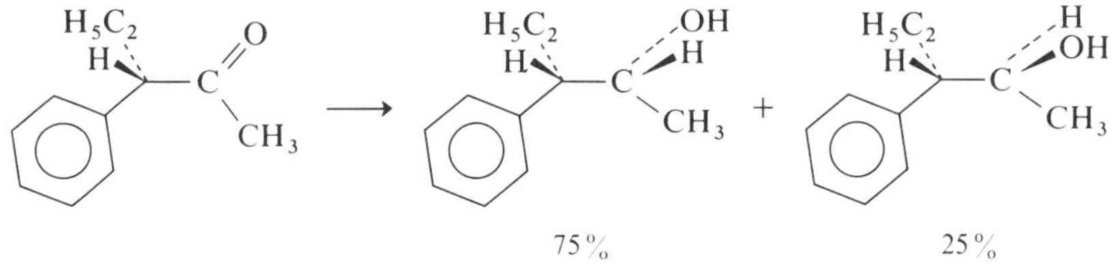

75% 25%

LITHIUM ALUMINUM HYDRIDE-ALUMINUM CHLORIDE REMOVAL OF OXYGEN

In positions where a highly stabilized carbonium ion intermediate can be anticipated, lithium aluminum hydride-aluminum chloride can be an effective reagent in removing oxygen. For example, a benzyl carbonium ion would be formed by loss of oxygen from a benzyl alcohol, and both diphenylcarbinol and acetophenone (which would be reduced to a benzyl alcohol) give hydrocarbons on treatment with lithium aluminum hydride-aluminum chloride. The mechanism probably proceeds by the following intermediates:

$$-\overset{\overset{\displaystyle O}{\|}}{C}- \longrightarrow -\overset{\overset{\displaystyle O-\overset{|}{Al}-}{|}}{CH}- \longrightarrow -\underset{+}{CH}- \longrightarrow -CH_2-.$$

Reduction of cholestenone also gives a hydrocarbon, via an intermediate allylic alcohol derivative.

$$C_6H_5-CH(OH)-C_6H_5 \xrightarrow[AlCl_3]{LiAlH_4} (C_6H_5)_2CH_2$$

$$C_6H_5-COCH_3 \xrightarrow[AlCl_3]{LiAlH_4} C_6H_5-CH_2CH_3$$

$$\text{cholestenone} \xrightarrow[AlCl_3]{LiAlH_4} \text{cholestene}$$

Treatment of an acetal with the reagent can also give a resonance-stabilized cation, $-\overset{\oplus}{C}H-OR \longleftrightarrow -CH=\overset{\oplus}{O}R$, which is reduced to an ether.

$$RCH(OEt)_2 \xrightarrow[AlCl_3]{LiAlH_4} RCH_2OEt$$

$$C_6H_5-\text{(1,3-dioxolan-2-yl)} \xrightarrow[AlCl_3]{LiAlH_4} C_6H_5-CH_2OCH_2CH_2OH$$

Epoxides also react with lithium aluminum hydride-aluminum chloride, not because of a stable carbonium ion intermediate but because of strain in the epoxide. Although epoxides are opened by lithium aluminum hydride alone, the favored product with the complex is different, being that predicted from opening to give the more stable carbonium ion.

$$R_2\overset{O}{\overset{/ \backslash}{C-CHR}} \xrightarrow[AlCl_3]{LiAlH_4} R_2CH-\overset{\overset{\displaystyle OH}{|}}{C}HR$$

OTHER HYDRIDE REDUCTIONS

Although lithium aluminum hydride is an exceedingly valuable reagent and is the hydride reducing agent one usually thinks of first, it does suffer from a number of disadvantages (principally its own reactivity). It reacts with nearly all carbonyl compounds; thus, it cannot be used as a highly selective reagent, e.g., for aldehydes and ketones. It also reacts with alcohols and with water much as a Grignard reagent does, liberating hydrogen. It is for this reason that lithium aluminum hydride reductions are usually carried out in ether.

Fortunately, a whole array of metal hydrides is available, varying in reactivity and specificity. Of the other hydrides, sodium borohydride, unlike lithium aluminum hydride, does not react rapidly with water or alcohol; thus, reductions can be carried out in these solvents. Sodium borohydride is, then, the reagent of choice for reduction of carbohydrates, which are insoluble in ethers. It is also the reagent of choice for the reduction of aldehydes and ketones in the presence of acids, esters, etc. The examples shown illustrate the use of sodium borohydride.

D-glucose $\xrightarrow[H_2O]{NaBH_4}$ sorbitol (CH_2OH–CH(OH)–CH(OH)... : $HOCH_2$–H–C–OH / HO–C–H / H–C–OH / H–C–OH–CH_2OH)

$$CH_3COCH_2COCH_3 \xrightarrow[H_2O\text{-}CH_3OH,\ 15^\circ]{NaBH_4} CH_3CH(OH)CH_2CH(OH)CH_3$$

90% meso, 2% racemic

3-oxo-1-indanylacetic acid (CH_2COOH, C=O) $\xrightarrow[NaOH,\ H_2O,\ 25^\circ]{NaBH_4}$ $\xrightarrow{H^+}$ 96% 3-hydroxy-1-indanylacetic acid (CH_2COOH, OH)

methyl 4-oxocyclohexanecarboxylate ($COOCH_3$) $\xrightarrow[CH_3OH,\ 0^\circ]{NaBH_4}$ methyl 4-hydroxycyclohexanecarboxylate (OH, $COOCH_3$)

Among the other hydrides available for selective reduction, some of the more important are lithium borohydride, lithium tri-tert-butoxyaluminum hydride, and diborane. These reagents are prepared according to the following equations.

$$KBH_4 + LiCl \xrightarrow{THF} LiBH_4$$

$$(CH_3)_3COH + LiAlH_4 \xrightarrow{60^\circ} LiAlH(OtBu)_3$$

$$NaBH_4 + AlCl_3 \longrightarrow B_2H_6$$

$$3NaBH_4 + 4BF_3 \xrightarrow{diglyme} 3NaBF_4 + 2B_2H_6$$

Table 6-2 compares the reactivity of these hydride reducing reagents with that of sodium borohydride. Lithium aluminum hydride reduces all the functional groups shown. From the table, lithium borohydride's reactivity is apparently intermediate between those of lithium aluminum hydride and sodium borohydride. More particularly, lithium borohydride is not as reactive as lithium aluminum hydride but it does reduce esters, whereas sodium borohydride does not. The reducing power of lithium tri-(*t*-butoxy)-aluminum hydride is similar to that of sodium borohydride, but it is also very useful in the reduction of acid derivatives to aldehydes (see p. 117). Diborane contrasts with sodium borohydride in that it reduces acids but not acid chlorides.

Table 6-2

SELECTIVITY OF METAL HYDRIDES IN REDUCING FUNCTIONAL GROUPS

Functional group	Hydride			
	$NaBH_4$	$LiBH_4$	$LiAlH(OtBu)_3$	B_2H_6
RCHO	+	+	+	+
RCOR	+	+	+	+
RCOOH	–	–	–	+ fast
RCOOR′	–	+	–	+ slow
RCOCl	+	+	+	–

A particularly interesting reduction is that of ethyl *p*-nitrobenzoate to *p*-nitrobenzyl alcohol, which can be carried out at room temperature with lithium borohydride. The comparable reduction with sodium borohydride does not go, whereas that with lithium aluminum hydride, unless carried out under closely defined conditions, gives reduction of the nitro group as well.

$$O_2N-C_6H_4-COOC_2H_5 \xrightarrow[20^\circ,\ 4\ hr]{LiBH_4} 88\%\ O_2N-C_6H_4-CH_2OH$$

CATALYTIC REDUCTION OF CARBONYL GROUPS

Prior to the introduction of metal hydrides as reducing agents, the normal method of reduction of a carbonyl group was catalytic reduction. In the laboratory, hydride reductions have now nearly eliminated catalytic reduction, since rather vigorous conditions are usually required to reduce carbon-oxygen double bonds catalytically. However, this method remains a useful procedure on an industrial scale, because hydrogen gas is quite cheap, whereas the metal hydrides are very expensive. For example, the reduction of triglycerides (present in fats and oils) to primary alcohols can be carried out over copper chromite catalyst. Other examples are shown. Most involve quite strenuous conditions, especially true for the acid derivatives.

$$\begin{array}{l} CH_2O-C(=O)(CH_2)_{16}CH_3 \\ CHO-C(=O)(CH_2)_{16}CH_3 \\ CH_2O-C(=O)(CH_2)_{16}CH_3 \end{array} + 6H_2 \xrightarrow[250^\circ]{CuCrO_4} \begin{array}{l} CH_2OH \\ CHOH \\ CH_2OH \end{array} + 3HOCH_2(CH_2)_{16}CH_3$$

$$\begin{array}{l} CHO \\ C(CH_3)_2 \\ CH_2N(C_2H_5)_2\cdot HCl \end{array} + \underset{40\ atm}{H_2} \xrightarrow[45^\circ]{Ni(R),\ H_2O} 90\%\ \begin{array}{l} CH_2OH \\ C(CH_3)_2 \\ CH_2N(C_2H_5)_2\cdot HCl \end{array}$$

$$\text{cyclopentanone} + \underset{70\ atm}{H_2} \xrightarrow[50^\circ]{Ni} \text{cyclopentanol (OH, H)}$$

$$\text{6-}CH_3O\text{-1-tetralone} + \underset{1\ atm}{H_2} \xrightarrow[r.t.]{PtO_2,\ EtOH} 97\%\ \text{6-}CH_3O\text{-1-tetralol}$$

$$HOCH_2COOH + \underset{400\ atm}{H_2} \xrightarrow[150^\circ,\ H_2O]{10\%\ Ru/C} 80\%\ HOCH_2CH_2OH$$

$$O{=}\text{pyrrolidine(NH)}-COOC_4H_9 + \underset{350\ atm}{2H_2} \xrightarrow[150^\circ]{CuCrO_4,\ dioxane} 95\%\ O{=}\text{pyrrolidine(NH)}-CH_2OH$$

$$\text{3,3,5-trimethylcyclohexanone} + \underset{1\ atm}{H_2} \xrightarrow[HOAc,\ 20^\circ]{Pt} \underset{83\%}{\text{alcohol (OH axial... H)}} + \underset{17\%}{\text{alcohol (H, OH)}}$$

An important generalization to note here is that hydrides normally reduce carbon-oxygen double bonds much easier than carbon-carbon double bonds, whereas the reverse is true for catalytic reduction. Thus, considerable latitude is allowed in the selective reductions of compounds containing both types of functional groups, as illustrated by reduction of cinnamaldehyde to cinnamyl alcohol with sodium borohydride and to β-phenylpropionaldehyde with hydrogen over palladium catalyst.

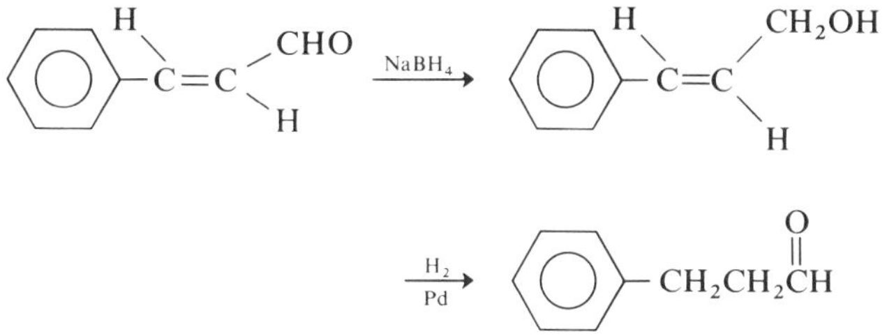

$$C_6H_5CH{=}CHCHO \xrightarrow{NaBH_4} C_6H_5CH{=}CHCH_2OH$$

$$C_6H_5CH{=}CHCHO \xrightarrow[Pd]{H_2} C_6H_5CH_2CH_2CHO$$

MEERWEIN-PONNDORF-VERLEY REDUCTION

When the Oppenauer oxidation was discussed in Chapter 5, it was noted that the reaction is reversible. In fact, the reverse reaction, the Meerwein-Ponndorf-Verley reduction, is just as useful as the Oppenauer oxidation. To shift the equilibrium, we employ an excess of the secondary alcohol reductant, normally isopropyl alcohol, together with an aluminum alkoxide catalyst, and the acetone produced, the most volatile constituent in the equilibrium, is distilled out as it is formed.

Since the Meerwein-Ponndorf reduction does involve an equilibrium, it seems reasonable that the most stable alcoholic product should be formed. This can be illustrated in many reactions, but nowhere better than in the reduction of steroids, where, for example, 3β-cholestanol is formed as the much preferred product of the reduction of cholestanone.

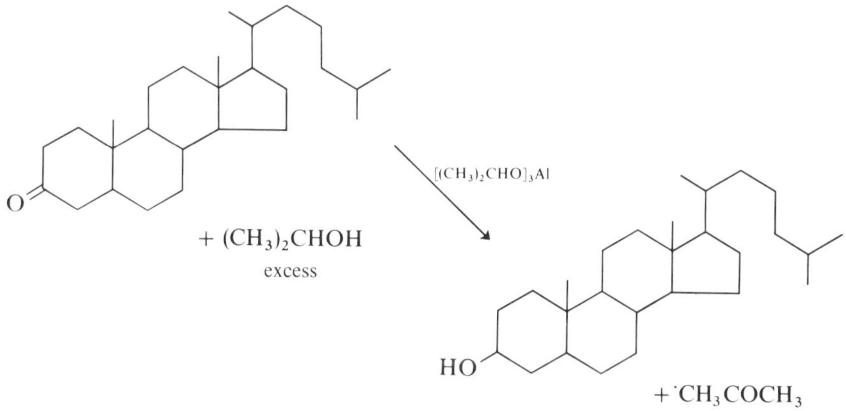

Cholestanone + $(CH_3)_2CHOH$ (excess) $\xrightarrow{[(CH_3)_2CHO]_3Al}$ 3β-cholestanol (HO) + CH_3COCH_3

The Meerwein-Ponndorf-Verley reduction is often as selective as hydride reduction. For example, a keto group can be reduced in the presence of a nitro group, and a conjugated ketone does not give alkene reduction. Use of the Meerwein-Ponndorf-Verley reduction is limited only by its strongly basic conditions.

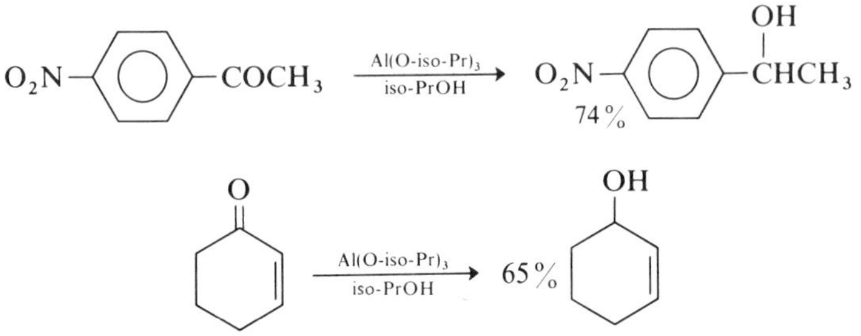

$O_2N-C_6H_4-COCH_3 \xrightarrow[\text{iso-PrOH}]{\text{Al(O-iso-Pr)}_3} O_2N-C_6H_4-CH(OH)CH_3$ 74%

Al(O-iso-Pr)$_3$, iso-PrOH → 65%

The Meerwein-Ponndorf-Verley reduction (and, therefore, the Oppenauer oxidation) involves a direct hydride transfer from the aluminum alkoxide to the carbonyl compound. In this way, an optically active secondary alcohol employed as reducing agent leads to an optically active secondary alcohol as product.

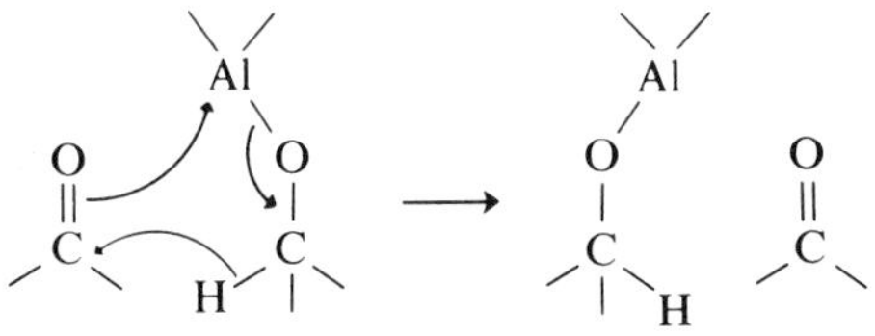

REDUCTION OF CARBONYL GROUPS BY METALS

An effective reducing agent for ketones is sodium metal in alcohol. Like the Meerwein-Ponndorf-Verley reduction, this leads to the more stable of the potential alcoholic products, indicating that an equilibrium is probably involved. Similar reduction of an ester leads to a primary alcohol.

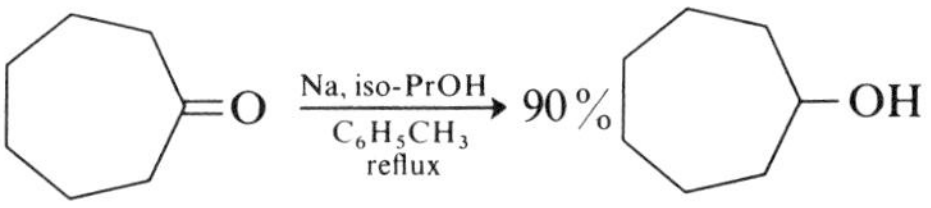

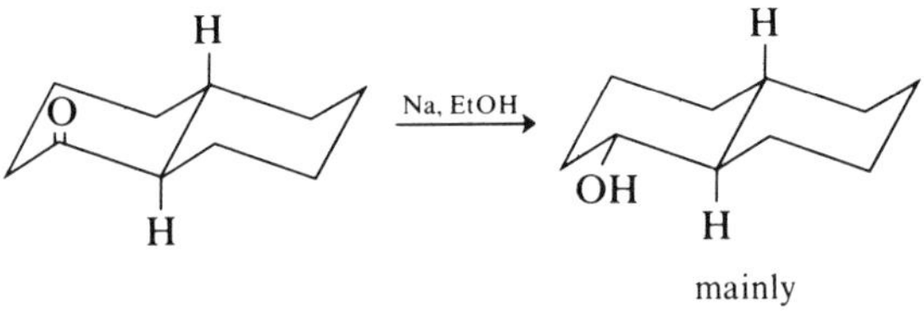

Na, iso-PrOH, $C_6H_5CH_3$, reflux → 90%

Na, EtOH →

mainly

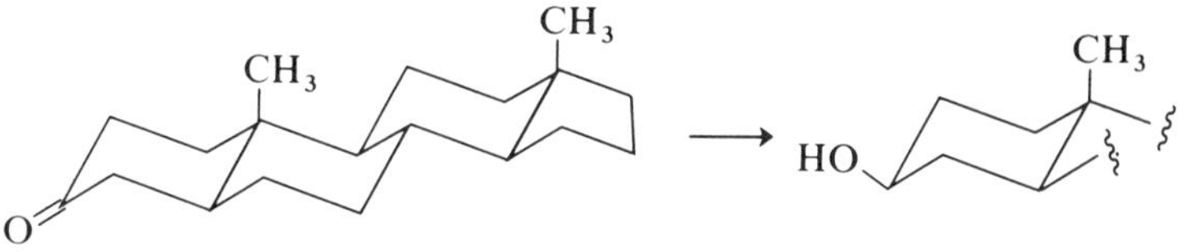

$$CH_3(CH_2)_{10}\overset{O}{\overset{\|}{C}}OC_2H_5 \xrightarrow[\substack{C_2H_5OH \\ C_6H_5CH_3 \\ \text{reflux}}]{Na} 75\%\ CH_3(CH_2)_{10}CH_2OH$$

Just as hydrogenation of ketones is less important now in the laboratory with the present availability of metal hydrides, so, too, sodium and alcohol reductions are no longer popular in the laboratory. Nevertheless, on an industrial scale, the reaction still is important, since metallic sodium is a relatively cheap reducing agent. The reduction of triglycerides to primary alcohols, useful in the preparation of detergents, is an important example.

$$\begin{array}{l} CH_2OCO(CH_2)_{16}CH_3 \\ | \\ CHOCO(CH_2)_{16}CH_3 \\ | \\ CH_2OCO(CH_2)_{16}CH_3 \end{array} \xrightarrow[C_2H_5OH]{Na} \begin{array}{l} CH_2OH \\ | \\ CHOH \\ | \\ CH_2OH \end{array} + 3HO(CH_2)_{17}CH_3$$

The mechanism of sodium metal reduction involves donation of an electron from a sodium atom to a carbonyl group, giving a radical anion, with the charge and free electron distributed on carbon and oxygen. This, in turn, picks up a proton from the solvent, giving an alkoxy radical. The radical is reduced by a second atom of sodium to an alkoxide anion, which on reaction with solvent gives the alcohol. A similar mechanism can be written for the reduction of an ester to a primary alcohol. Reduction to the alcohol is in both cases dependent on the availability of a good proton source in the solvent.

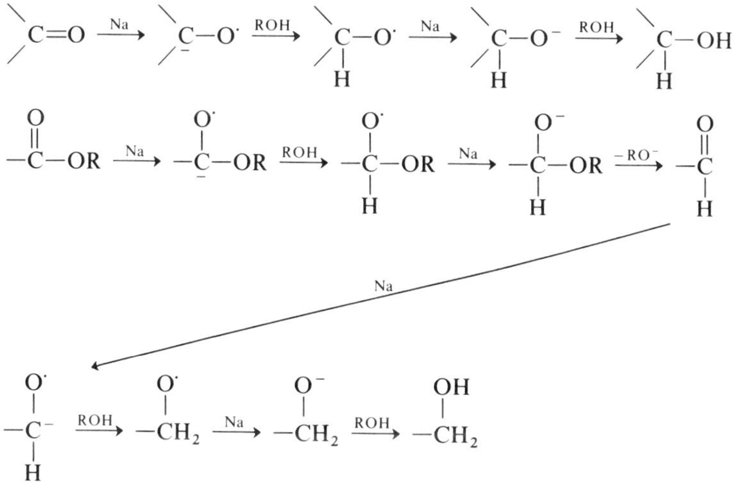

$$\gt C{=}O \xrightarrow{Na} \gt \overset{-}{C}{-}O^{\cdot} \xrightarrow{ROH} \gt CH{-}O^{\cdot} \xrightarrow{Na} \gt CH{-}O^{-} \xrightarrow{ROH} \gt CH{-}OH$$

$$-\overset{O}{\overset{\|}{C}}-OR \xrightarrow{Na} -\overset{O^{\cdot}}{\overset{|}{\underset{-}{C}}}-OR \xrightarrow{ROH} -\overset{O^{\cdot}}{\overset{|}{\underset{H}{\underset{|}{C}}}}-OR \xrightarrow{Na} -\overset{O^{-}}{\overset{|}{\underset{H}{\underset{|}{C}}}}-OR \xrightarrow{-RO^{-}} -\overset{O}{\overset{\|}{\underset{H}{\underset{|}{C}}}}$$

$$\xrightarrow{Na} -\overset{O^{\cdot}}{\overset{|}{\underset{H}{\underset{|}{C^{-}}}}} \xrightarrow{ROH} -\overset{O^{\cdot}}{\overset{|}{CH_2}} \xrightarrow{Na} -\overset{O^{-}}{\overset{|}{CH_2}} \xrightarrow{ROH} -\overset{OH}{\overset{|}{CH_2}}$$

An interesting point concerns the fate of the radical anion in the absence of a proton source. To these radical anions, like other radicals formed in reasonably high concentration, the simplest route open is coupling, and that reaction takes place. Benzpinacol, for example, is formed by sodium reduction of benzophenone. A similar reduction with magnesium metal amalgam gives pinacol from acetone.

$$C_6H_5COC_6H_5 \xrightarrow[\text{Et}_2\text{O or NH}_3\ (\text{liq})]{\text{Na}} (C_6H_5)_2C(OH){-}C(OH)(C_6H_5)_2$$

$$CH_3\overset{O}{\overset{\|}{C}}CH_3 \xrightarrow[\substack{C_6H_6 \\ \text{reflux}}]{\text{Mg(Hg)}} 50\%\ CH_3{-}C(CH_3)(OH){-}C(CH_3)(OH){-}CH_3$$

$$CH_3{-}\dot{C}(CH_3){-}O^- \longrightarrow (CH_3{-}C(CH_3)(O^-){+}_2Mg^{2+}$$

Sodium reduction of esters in the absence of a proton source gives α-hydroxy ketones. The radical anion from the ester couples to give an intermediate α-diketone. The diketone is then reduced further by metallic sodium to give, successively, the radical anion and dianion. Since the dianion is stable in the absence of a proton donor, the reaction stops there. Acidification of the dianion gives the enediol, which can tautomerize rapidly to an α-hydroxyketone. The sodium metal-catalyzed coupling of an ester, known as the acyloin reaction, is very useful. The reaction has proved especially effective for the formation of medium-sized rings (those containing 8 through 10 atoms) from diesters under conditions of very high dilution. It is, in fact, one of the few routes available for the formation of medium-sized rings.

$$CH_3(CH_2)_2\overset{O}{\overset{\|}{C}}OC_2H_5 \xrightarrow[\substack{(C_2H_5)_2O \\ \text{reflux}}]{\text{Na}} 70\%\ CH_3(CH_2)_2\overset{O}{\overset{\|}{C}}\overset{OH}{\overset{|}{C}}H(CH_2)_2CH_3$$

$$-\overset{O}{\overset{\|}{C}}OC_2H_5 \xrightarrow{\text{Na}} -\overset{O^-}{\overset{|}{\dot{C}}}{-}OC_2H_5 \longrightarrow -C(O^-)(OC_2H_5){-}C(O^-)(OC_2H_5)- \xrightarrow{-2C_2H_5O^-} -\overset{O}{\overset{\|}{C}}{-}\overset{O}{\overset{\|}{C}}-$$

$$-\overset{O}{\overset{\|}{C}}{-}\overset{O}{\overset{\|}{C}}- \xrightarrow{\text{Na}} -\overset{O^-}{\overset{|}{C}}{=}\overset{O^\cdot}{\overset{|}{C}}- \xrightarrow{\text{Na}} -\overset{O^-}{\overset{|}{C}}{=}\overset{O^-}{\overset{|}{C}}- \xrightarrow{\text{ROH}} -\overset{OH}{\overset{|}{C}}{=}\overset{OH}{\overset{|}{C}}- \longrightarrow -\overset{O}{\overset{\|}{C}}{-}\overset{OH}{\overset{|}{C}}H-$$

$$\begin{array}{c} CO_2CH_3 \\ | \\ (CH_2)_8 \\ | \\ CO_2CH_3 \end{array} \xrightarrow[\substack{\text{reflux} \\ \text{high} \\ \text{dilution}}]{Na \atop C_6H_4(CH_3)_2} \xrightarrow{HOAc} 70\%$$ (2-hydroxycyclodecanone: cyclic ketone with =O and adjacent OH)

REFERENCES

Augustine, R. L., *Catalytic Hydrogenation*, Marcel Dekker, Inc., New York, 1965.

Augustine (*Reduction*): Chap. 1, M. N. Rerick, "The Chemistry of the Mixed Hydrides."

Augustine (*Reduction*): Chap. 2, M. Smith, "Dissolving Metal Reductions."

Bentley (Part 1): Chap. 9, F. J. McQuillin, "Reduction and Hydrogenation in Structural Elucidation."

Finley, K. T., "The Acyloin Condensation as a Cyclization Method," *Chem. Rev.*, **64**, 573 (1964).

Foerst (Vol. 4): pp. 196–335, E. Schenker, "The Use of Complex Borohydrides and of Diborane in Organic Chemistry."

Gaylord, N. G., "Reduction with Complex Metal Hydrides," Interscience Publishers, New York, 1956.

House: Chap. 2, "Metal Hydride Reductions and Related Reactions."

Rylander, P. N., "Catalytic Hydrogenation over Platinum Metals," Academic Press, Inc., New York, 1967, Part IV ("Carbonyl Compounds").

Smith, H., "Hydrogenation of Unsaturated Systems with Metal-Ammonia Reagents," in *Organic Reactions in Liquid Ammonia*, John Wiley & Sons, Inc., New York, 1963, pp. 212–78.

Wheeler, O. H., "Reduction of Carbonyl Groups," in *Chemistry of the Carbonyl Group*, S. Patai, ed., John Wiley & Sons, Inc., New York, 1966, Chap. 11.

Wilds, A. L., "Reduction with Aluminum Alkoxides," *Org. Reactions*, **2**, 178 (1944).

PROBLEMS

1. Suggest reaction conditions (reagents, catalysts, solvents, etc.) suitable for effecting the following conversions. More than one step may be required.

(a) $O_2N-C_6H_4-COOC_2H_5 \longrightarrow O_2N-C_6H_4-CH_2OH$

(b) 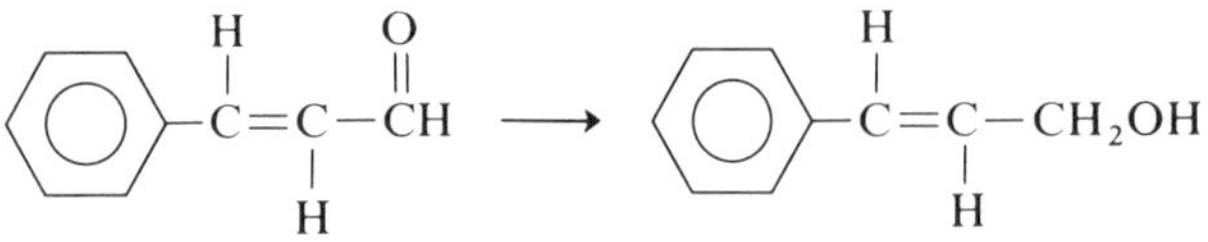 $C_6H_5\text{-}CH{=}CH\text{-}CHO \longrightarrow C_6H_5\text{-}CH{=}CH\text{-}CH_2OH$

(c) $CH_3CH{=}CHCH_2COCH_3 \longrightarrow CH_3CH{=}CHCH_2CH(OH)CH_3$

(d) 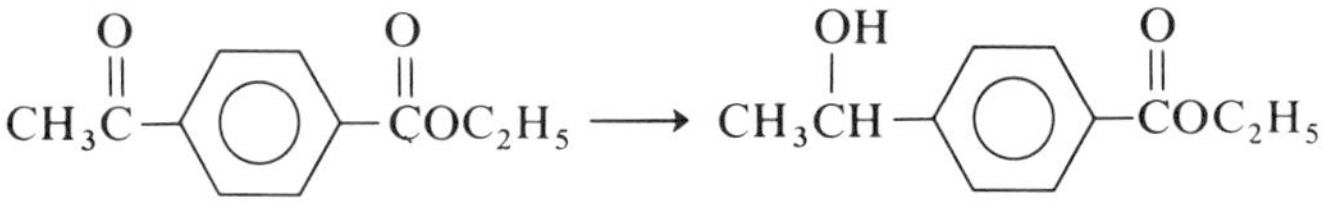 $CH_3CO\text{-}C_6H_4\text{-}COOC_2H_5 \longrightarrow CH_3CH(OH)\text{-}C_6H_4\text{-}COOC_2H_5$ (para)

(e) 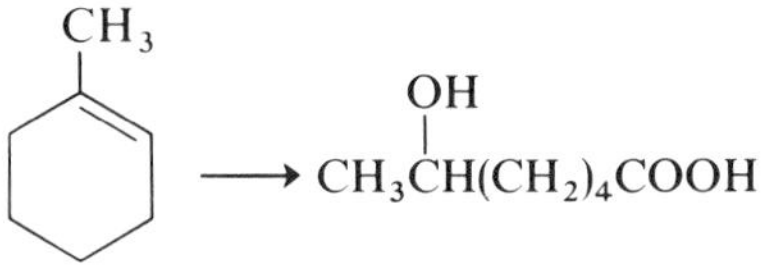 1-methylcyclohexene (ring with CH_3) $\longrightarrow CH_3CH(OH)(CH_2)_4COOH$

(f) 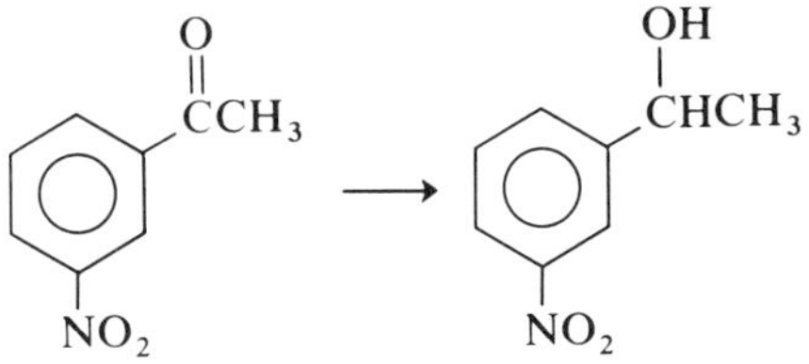 $m\text{-}O_2N\text{-}C_6H_4\text{-}COCH_3 \longrightarrow m\text{-}O_2N\text{-}C_6H_4\text{-}CH(OH)CH_3$

(g) $C_2H_5OOC(CH_2)_8COOC_2H_5 \longrightarrow$ cyclodecanone ring bearing OH on the carbon next to $C{=}O$

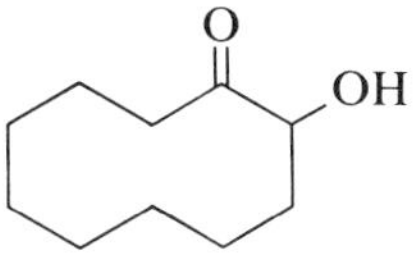

2. Indicate the expected product of the following reaction including stereochemistry.

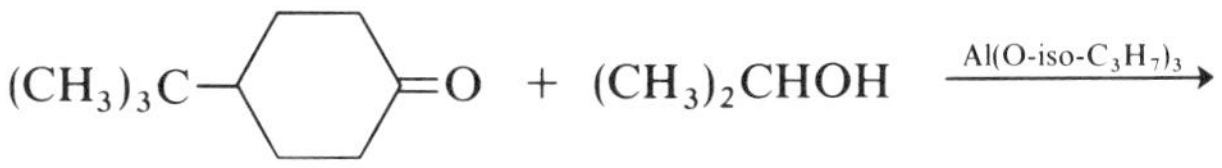

$$(CH_3)_3C\text{-}C_6H_{9}{=}O + (CH_3)_2CHOH \xrightarrow{Al(O\text{-}iso\text{-}C_3H_7)_3}$$

(4-*tert*-butylcyclohexanone)

3. Write balanced equations for the reduction steps of Problems 1(a), 1(c), and 2.

4. On page 105, reduction of a ketone with sodium is shown to give the intermediate $-\overset{|}{\underset{-}{C}}-O^{\cdot}$, whereas on page 106, reduction of a ketone with magnesium is indicated to give $-\overset{|}{\underset{\cdot}{C}}-O^{-}$. Are the two radical anions different? Explain.

7

Interconversion of Aldehydes and Ketones with Acid Derivatives

OXIDATION OF ALDEHYDES TO ACIDS

In Chapter 5, we noted that it is difficult to oxidize a primary alcohol to an aldehyde, since the aldehyde itself is more susceptible to oxidation than the alcohol. It is not surprising, then, that aldehydes can be oxidized readily on a preparative scale to the corresponding carboxylic acids in excellent yields. Examples of the oxidation of aldehydes by relatively vigorous reagents follow. The last compound oxidized, formic acid, is not really an aldehyde, but it contains the HCO— (formyl) group common to aldehydes.

$$CH_3(CH_2)_5\overset{\overset{O}{\|}}{C}H \xrightarrow[\text{aq } H_2SO_4,\ 20^\circ]{KMnO_4} 78\%\ CH_3(CH_2)_5COOH$$

$$\text{3,4-}(CH_2O_2)C_6H_3\text{—}CHO \xrightarrow[H_2O,\ 75^\circ]{KMnO_4} 84\%\ \text{3,4-}(CH_2O_2)C_6H_3\text{—}COOH$$

$$\text{2-furyl—}CHO \xrightarrow{H_2CrO_4} 75\%\ \text{2-furyl—}COOH$$

$$ClCH_2CH_2\overset{\overset{O}{\|}}{C}H \xrightarrow[33^\circ]{\text{fuming } HNO_3} 65\%\ ClCH_2CH_2COOH$$

$$HCOOH \xrightarrow{KMnO_4} CO_2 + H_2O$$

The oxidation of an aldehyde to a carboxylic acid is so easy that it can proceed under considerably milder conditions than those we have just cited, and the selective oxidation of an aldehyde in the presence of other oxidizing groups is often employed. One reagent that selectively oxidizes water-soluble aldehydes (especially sugars) is bromine water with strontium carbonate added as a base. The base is apparently necessary to convert the hemiacetal to its alkoxide, which has been shown to be the species reacting with bromine.

$$\text{D-glucose} \xrightarrow[SrCO_3]{Br_2,\ H_2O} \text{D-gluconolactone}$$

$$\text{D-glucose} \xrightarrow{CO_3^{2-}} {-O}(O^-)C(H) \xrightarrow{Br_2} {-O}C{=}O + HBr + Br^-$$

In addition to those reagents that can convert aldehydes preparatively to acids are a number of metal ions which aldehydes reduce to a lower oxidation state, thus constituting a qualitative test for the presence of the aldehydes. In these cases, the reduction products of the metal ions are observed but the organic products are rarely isolated. One of these qualitative tests that can also be employed on a preparative scale is the Tollens test. In the test with Tollens reagent, silver ion is reduced to silver metal, observable either as a black precipitate or a silver mirror. For this reason, the Tollens test is sometimes referred to as the silver mirror test. The Tollens reagent is effective for nearly all aldehydes, including those containing other very sensitive functional groups or rings.

$$\text{2-furaldehyde (CHO)} \xrightarrow[AgO/Cu_2O]{O_2,\ NaOH} \text{2-furoic acid (COOH)} + Ag^0$$

$$\text{thiophene-3-carbaldehyde (CHO)} \xrightarrow[\substack{NaOH,\ H_2O \\ 0^\circ}]{Ag_2O} 97\%\ \text{thiophene-3-carboxylic acid (COOH)} + Ag^0$$

$$\text{4-hydroxy-3-methoxybenzaldehyde (OH, } OCH_3\text{, CHO)} \xrightarrow[NaOH]{Ag_2O} \text{(ONa, } OCH_3\text{, } CO_2Na\text{)} + Ag^0$$

Cupric ion, the active ingredient in both Benedict's and Fehling's tests, is reduced to cuprous oxide (red, pink, or yellow) by aldehydes with α-hydroxyl groups.

$$\text{D-glucose} + Cu^{2+} \xrightarrow{\text{buffer}} Cu_2O\downarrow$$

D-glucose

$$\text{D-fructose} + Cu^{2+} \xrightarrow{\text{buffer}} Cu_2O\downarrow$$

D-fructose

The reaction is not a specific test for the aldehyde group, since only α-hydroxy aldehydes are oxidized. Moreover, it is not restricted to aldehydes because α-hydroxy ketones also give positive tests. It does constitute a test for sugars and can, in fact, be employed as a quantitative estimate of the amount of sugar in blood.

DISPROPORTIONATION OF ALDEHYDES TO ACIDS AND ALCOHOLS (THE CANNIZZARO REACTION)

A curious reaction involving simultaneous oxidation and reduction (disproportionation) of an aldehyde, to give an acid and an alcohol, is the Cannizzaro reaction, effected by heating the aldehyde in concentrated aqueous hydroxide. Since these are precisely the conditions employed for the aldol condensation, the only aldehydes that undergo the reaction are those without α-hydrogens, i.e., those like benzaldehyde, formaldehyde, and trimethylacetaldehyde (pivaldehyde). Yields of alcohols and acids are usually quite good from such aldehydes, as can be seen from the illustration.

$$2\ o\text{-}NO_2C_6H_4CHO \longrightarrow 91\%\ o\text{-}NO_2C_6H_4CH_2OH + 91\%\ o\text{-}NO_2C_6H_4COOH$$

$$2(CH_3)_3CCHO \longrightarrow (CH_3)_3CCH_2OH + (CH_3)_3CCOOH$$

The Cannizzaro reaction is first order in hydroxide ion. Its mechanism involves an intermolecular hydride ion transfer, from one benzaldehyde molecule to another. This has been shown by the use of deuterium-labeled benzaldehyde, a reaction leading to benzyl alcohol containing two deuterium atoms in its carbinol group.

$$2\ C_6H_5CHO \xrightarrow{NaOH} C_6H_5CH(OH)O^-Na^+ + C_6H_5CHO \longrightarrow$$

$$C_6H_5COOH + {}^+Na^-OCH_2C_6H_5 \longrightarrow C_6H_5COO^-Na^+ + HOCH_2C_6H_5$$

$$C_6H_5CDO \xrightarrow{NaOH} C_6H_5CD_2OH + C_6H_5COONa$$

A number of variations of the Cannizzaro reaction have been employed. One is an intramolecular Cannizzaro reaction, leading to a hydroxy acid; for instance, glyoxal gives glycolic acid. A related reaction involves phenyl group transfer, in which benzil gives benzilic acid (in the so-called benzilic acid rearrangement).

$$\underset{\text{glyoxal}}{HCOCHO} \xrightarrow{NaOH} \underset{\text{glycolic acid}}{HOCH_2COOH}$$

$$\underset{\text{benzil}}{C_6H_5CO-COC_6H_5} \xrightarrow{NaOH} \underset{\text{benzilic acid}}{(C_6H_5)_2C(OH)-COOH}$$

$$\downarrow OH^-$$

$$C_6H_5C(=O)-C(O^-)(OH)C_6H_5 \longrightarrow (C_6H_5)_2C(O^-)-C(=O)OH$$

Formaldehyde is exceptionally good as a reducing agent and sometimes is employed with other aldehydes to reduce them to alcohols in what has come to be called a crossed Cannizzaro reaction. Another variation of the Cannizzaro reaction leads to pentaerythritol. First, acetaldehyde reacts three times in aldol reactions with formaldehyde; these are followed by a crossed Cannizzaro reaction employing excess formaldehyde.

$$ArCH(=O) + CH_2O \xrightarrow{NaOH} ArCH_2OH + HCOOH$$

$$CH_3CH(=O) + 3CH_2O \xrightarrow{Ca(OH)_2} (HOCH_2)_3CCH(=O) \xrightarrow{CH_2O} 57\%\ (HOCH_2)_4C + HCOOH$$

Related to the Cannizzaro reaction is the Tischchenko reaction, where alkoxide ion is the basic catalyst, instead of hydroxide ion. The mechanism is presumably the same, involving hydride ion transfer, but here the oxidation product isolated is an ester rather than an acid. The Tischchenko reaction is employed commercially in the preparation of ethyl acetate from acetaldehyde. Its mechanism is illustrated for the formation of benzyl benzoate from benzaldehyde.

$$2CH_3CH(=O) \xrightarrow{Al(OC_2H_5)_3} CH_3CO{-}OCH_2CH_3$$

$$2\ C_6H_5CH(=O) \xrightarrow[\text{catalyst}]{C_6H_5CH_2ONa} C_6H_5CO{-}OCH_2C_6H_5$$

$$\downarrow C_6H_5CH_2O^-\ Na^+$$

$$C_6H_5C(O^-)(H)(OCH_2C_6H_5) \quad Na^+ \quad C_6H_5CH(=O) \longrightarrow C_6H_5CO{-}OCH_2C_6H_5 + Na^+\ {}^-O{-}CH_2C_6H_5$$

BAEYER-VILLIGER OXIDATION OF KETONES

Like most other functional groups in aliphatic molecules, ketones can be oxidized to acids by vigorous oxidizing agents in a rather nonselective fashion. Use is made of this reaction, for example, in the industrial synthesis of adipic acid (a 6,6-nylon precursor) from mixed cyclohexanone and cyclohexanol.

$$\text{cyclohexanone} \left(+ \text{cyclohexanol} \right) \xrightarrow[100^\circ]{HNO_3} HOOC(CH_2)_4COOH + NO + H_2O$$

Ketones can also be oxidized rather selectively, by peracids to esters in the Baeyer-Villiger oxidation. Typical conditions are found in the two reactions shown.

$$CH_3COC_6H_5 \xrightarrow[25^\circ,\ CHCl_3,\ 10\ days]{C_6H_5CO_3H} 63\%\ CH_3C(=O)-O-C_6H_5$$

$$CH_3CO\text{-cyclopropyl} \xrightarrow[13^\circ,\ CF_3COOH,\ Na_2HPO_4]{CF_3CO_3H} 53\%\ CH_3C(=O)-O\text{-cyclopropyl}$$

Points which must be explained by any mechanism postulated for the Baeyer-Villiger reaction are that the reaction is acid catalyzed but cannot involve a free carbonium ion; that it is faster for peracids substituted by electronegative groups and for ketones bearing electropositive substituents; that tertiary alkyl groups migrate faster than aryl groups, but aryl groups migrate faster than methyl groups; and that the configuration of the starting material is retained.

$$\text{bicyclic}-COC_6H_5 \xrightarrow[HOAc]{CF_3CO_3H} \text{good yield}\ \text{bicyclic}-OCOC_6H_5$$

$$O_2N-C_6H_4-CO-C_6H_5 \xrightarrow[HOAc,\ H_2SO_4,\ 25^\circ]{CH_3CO_3H} O_2N-C_6H_4-CO-O-C_6H_5$$

95%

$$\text{(norbornyl)}-COCH_3 \xrightarrow[CHCl_3,\ 25°]{C_6H_5CO_3H} 81\%\ \text{(norbornyl)}-OCOCH_3$$

optically active

The mechanism shown would explain those observations very well.

$$Ar\overset{^{18}O}{\overset{\|}{C}}CH_3 \xrightarrow[H_2SO_5]{RCO_3H \text{ or}} Ar\overset{^{18}OH}{\overset{|}{\underset{O_3CR}{\underset{|}{C}}}}CH_3 \longrightarrow \left[Ar \cdots C(^{18}O\text{-}H)(-CH_3) \cdots O\text{-}O_2CR \right] \longrightarrow Ar{-}O{-}\overset{^{18}O}{\overset{\|}{C}}{-}CH_3 + H^+ + {}^-O_2CR$$

Studies employing ^{18}O show that the carbonyl oxygen of the ketones becomes the carbonyl oxygen of the ester, in accord with the reaction mechanism shown.

A number of modifications of the Baeyer-Villiger oxidation are known. Oxidation of a cyclic ketone gives, predictably, a lactone, whereas oxidation of an α-diketone gives an anhydride. The latter reaction may also be carried out with hydrogen peroxide in base, in which event 2 moles of acid are formed. When the organic reactant is an α-keto acid, the acid one carbon atom shorter is the product, plus carbon dioxide.

$$\text{cyclopentanone} \xrightarrow[CF_3COOH]{CF_3CO_3H} 81\%\ \text{δ-valerolactone}$$

$$CH_3COCOCH_3 \xrightarrow{RCO_3H} CH_3CO{-}O{-}COCH_3$$

$$R{-}\overset{O}{\overset{\|}{C}}{-}\overset{O}{\overset{\|}{C}}{-}R \xrightarrow[OH^-]{H_2O_2} 2RCOOH$$

$$R{-}\overset{O}{\overset{\|}{C}}{-}COOH \longrightarrow RCOOH + H_2O + CO_2$$

A reaction related to the Baeyer-Villiger reaction is the Dakin reaction, in which an aromatic aldehyde or ketone is converted to the corresponding phenol. This reaction is limited to *ortho*- and *para*-hydroxy aromatic aldehydes and ketones, however. The mechanism shown would explain this specificity, since the transition state approximating an intermediate benzenonium ion can be stabilized by hydroxyl substitution in ortho or para positions.

$$\text{o-}HOC_6H_4CHO \xrightarrow{H_2O_2,\ OH^-} 73\%\ \text{o-}C_6H_4(OH)_2 + HCOOH$$

$$\text{o-}HOC_6H_4CHO \longrightarrow \text{o-}HOC_6H_4CH(OH)OOH \longrightarrow \left[\overset{\delta^-}{HO}\cdots O\text{-}CH(OH)\cdots\overset{\delta^+}{C_6H_4OH} \right] \longrightarrow \text{o-}HOC_6H_4OCHO + H_2O$$

HYDRIDE REDUCTION OF ACID DERIVATIVES TO ALDEHYDES

In the usual sequences of synthetic organic chemistry, acids are relatively easy to obtain; aldehydes are much more difficult. Thus, on a synthetic basis it is more helpful to reduce an acid derivative to an aldehyde then the reverse. For this reason, a number of special methods have been developed. No single procedure appears to give good yields for all types of compounds, and the decision as to which reagent to employ is often a matter of trial and error. Usually the reagents employed are deactivated forms of lithium aluminum hydride, such as lithium di- or tri-alkoxy aluminum hydrides. Since the mechanism of reduction involves initial addition of hydride ion to the carbonyl group, the electronegative alkoxide groups lower the electron density on the hydride ion, rendering it a less active, more selective, reagent.

$$-\overset{-}{Al}-H \quad X-\overset{O}{\overset{\|}{C}}-R \longrightarrow -Al\langle \quad H-\overset{O^-}{\overset{|}{C}}(X)-R \longrightarrow H-\overset{O}{\overset{\|}{C}}-R + X^-$$

The preferred acid derivatives for reduction appear to be tertiary amides; secondary amides are unsuccessful due to their acidic properties. Other acid derivatives—nitriles and acid chlorides, even especially active esters—have also been used in some cases.

$$\text{C}_6\text{H}_{11}-\text{CO}-\text{N(CH}_3)_2 \xrightarrow[\text{Et}_2\text{O}, 0°]{\text{LiAlH}_2(\text{OEt})_2} 71\%\ \text{C}_6\text{H}_{11}-\text{CHO}$$

$$\text{CH}_3(\text{CH}_2)_2\text{C}\equiv\text{N} \xrightarrow[\text{Et}_2\text{O}]{\text{LiAlH}_2(\text{OEt})_2} 68\%\ \text{CH}_3(\text{CH}_2)_2\text{CHO}$$

$$\text{CH}_3(\text{CH}_2)_2\text{C}\equiv\text{N} \xrightarrow{\text{HAl(iso-Bu)}_2} 93\%\ \text{CH}_3(\text{CH}_2)_2\text{CHO}$$

$$O_2N-C_6H_4-\text{COCl} \xrightarrow[\text{diglyme}, -78°]{1\text{ eq LiAlH(OtBu)}_3} 81\%\ O_2N-C_6H_4-\text{CHO}$$

$$\text{RCOO}-C_6H_5 \xrightarrow{\text{LiAlH(OtBu)}_3} \text{RCHO}$$

With some tertiary amides, lithium aluminum hydride itself can often be employed, whereas more exotic hydrides (sodium trimethoxyborohydride, sodium aluminum hydride) sometimes are used for other derivatives. From the examples given, we see that the arsenal of organic chemistry contains many weapons for effecting this reduction.

$$\text{RCO}-\text{N(imidazolyl)} \xrightarrow{\text{LiAlH}_4} \text{RCHO}$$

$$\text{RCO}-\text{N(pyrazolyl)} \xrightarrow{\text{LiAlH}_4} \text{RCHO}$$

$$\text{C}_3\text{H}_5-\text{CO}-\text{N(aziridinyl)} \xrightarrow[\text{Et}_2\text{O}, 0°]{\text{LiAlH}_4} 60\%\ \text{C}_3\text{H}_5-\text{CHO}$$

$$C_6H_4(\text{CON(CH}_3)_2)_2 \xrightarrow[\text{THF}, 20°]{\text{LiAlD}_4} 38\%\ C_6H_4(\text{CDO})_2$$

40–50°

$$\text{COCl-cyclohexene(OAc)}_3 \xrightarrow{\text{NaBH(OCH}_3)_3} 85\%\ \text{CHO-cyclohexene(OAc)}_3$$

$$\text{CH}_3(\text{CH}_2)_2\text{COOCH}_3 \xrightarrow[-50°]{\text{NaAlH}_4} 81\%\ \text{CH}_3(\text{CH}_2)_2\text{CHO}$$

OTHER SYNTHESES OF ALDEHYDES FROM ACID DERIVATIVES

Prior to the advent of metal hydrides, a number of other reagents had been developed for converting various acid derivatives to aldehydes. In general, hydride reduction provides such a versatile array of possibilities that it is now preferred, but one or another of the older reductions can still find some use if the required derivative is available. Probably the most generally successful reduction of an acid derivative to an aldehyde is the Rosenmund reduction of an acid chloride over palladium-barium sulfate in the presence of a specially prepared catalyst poison (from quinoline), which slows the reduction of the aldehyde to the alcohol.

$$CH_3(CH_2)_9\underset{\substack{|\\CH_3}}{C}HCOCl + H_2 \xrightarrow[\text{quinoline}]{Pd/BaSO_4} 78\%\ CH_3(CH_2)_9\underset{\substack{|\\CH_3}}{C}HCHO$$
$$\text{1 atm}$$

The McFadyen-Stevens reduction proceeds by treatment of an acyl derivative of a sulfonylhydrazide with base. Due to the strong base involved, use of this reaction is limited to the preparation of aldehydes lacking α-hydrogens (aromatic or α-trisubstituted).

$$RCONHNHSO_2Ar \xrightarrow{OH^-} RCHO$$

The Stephen reduction involves formation of a chloroimine from a nitrile and its reduction by stannous chloride to an imine, which is then hydrolyzed to the aldehyde. The reaction is successful with both aromatic and aliphatic nitriles but is obviously more difficult to carry out than hydride reduction.

$$\text{(2-naphthyl)CN} \xrightarrow{HCl} -\underset{\substack{|\\Cl}}{C}{=}NH \xrightarrow[HCl]{SnCl_2} -CH{=}NH + SnCl_4 \xrightarrow{H_2O} 80\%\ \text{(2-naphthyl)CHO}$$

The Sonn-Müller reaction is closely related to the Stephen reaction in that a chloroimine is employed and stannous chloride is the reducing agent. However, in this reaction, the chloroimine is generated by treatment of a secondary amide with phosphorus pentachloride. The reaction is effectively limited to aromatic amides.

$$\text{(2-CH}_3\text{C}_6\text{H}_4)CONHC_6H_5 \xrightarrow{PCl_5} -\underset{\substack{|\\Cl}}{C}{=}N- \xrightarrow{SnCl_2} -CH{=}N- \xrightarrow{H_2O} 70\%\ \text{(2-CH}_3\text{C}_6\text{H}_4)CHO$$

REFERENCES

Augustine (*Oxidation*): Chap. 5, S. N. Lewis, "Peracid and Peroxide Oxidations."

Augustine (*Reduction*): Chap. 1, M. N. Rerick, "The Chemistry of the Mixed Hydrides."

Foerst (Vol. 4): pp. 209–44, E. Schenker, "The Use of Complex Borohydrides and of Diborane in Organic Chemistry."

Fuson, R. C., "Formation of Aldehydes and Ketones from Carboxylic Acids and Their Derivatives," in *Chemistry of the Carbonyl Group*, S. Patai, ed., John Wiley & Sons, Inc., New York, 1966, Chap. 4.

House: Chap. 2, "Metal Hydride Reductions and Related Reactions."

Lee, J. B., and B. C. Uff, "The Baeyer-Villiger Reaction," *Quart. Rev.* (London), **21**, 449 (1967).

Roček, J., "Oxidation of Aldehydes by Transition Metal Compounds," in *Chemistry of the Carbonyl Group*, S. Patai, ed., John Wiley & Sons, Inc., New York, 1966, Chap. 10.

Rylander, P. N., "Catalytic Hydrogenation over Platinum Metals," Academic Press, Inc., New York, 1967, Chap. 23 ("Hydrogenolysis of Acid Chlorides-Rosenmund Reduction").

Selman, S., and J. F. Eastham, "Benzilic Acid and Related Rearrangements," *Quart. Rev.* (London), **14**, 221 (1960).

Waters: Chap. 6, "Oxidation of Aldehydes."

Waters: Chap. 7, "The Oxidation of Ketones, Related Compounds and Carboxylic Acids."

Wiberg: Chap. 1, R. Stewart, "Oxidation by Permanganate."

Wiberg: Chap. 2, K. B. Wiberg, "Oxidation by Chromic Acid and Chromyl Compounds."

PROBLEMS

1. Write a detailed mechanism for each of the following reactions, including all intermediates and/or transition states.

(a) $2\,C_6H_5CHO \xrightarrow[\Delta]{NaOH} C_6H_5CH_2OH + C_6H_5COONa$

(b) $CH_3CHO + 3CH_2O \xrightarrow{Ca(OH)_2} (HOCH_2)_4C + HCOOH$

2. Suggest reaction conditions (reagents, catalysts, solvents, etc.) suitable for effecting the following conversion. More than one step may be required.

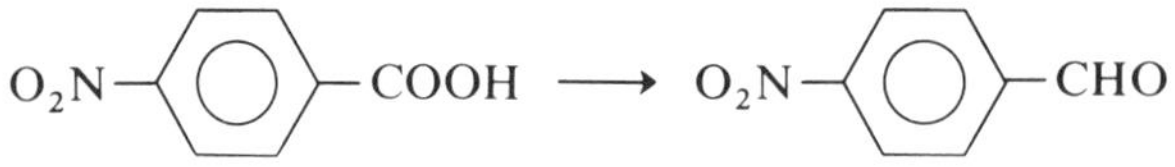

$$O_2N{-}C_6H_4{-}COOH \longrightarrow O_2N{-}C_6H_4{-}CHO$$

3. Indicate all the expected products of the following reactions. Show stereochemistry where appropriate.

(a) 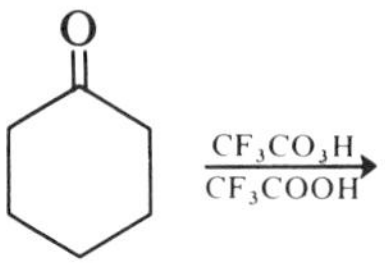 cyclohexanone $\xrightarrow[CF_3COOH]{CF_3CO_3H}$

(b) $C_6H_{11}{-}\overset{O}{\overset{\|}{C}}{-}N(CH_3)_2 \xrightarrow[0°]{LiAlH_2(OEt)_2}$

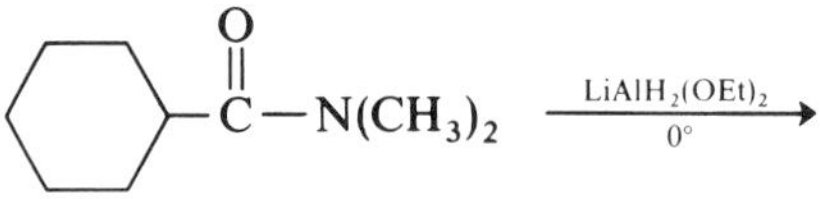

(c) phenanthrenequinone $\xrightarrow[\Delta]{NaOH}$

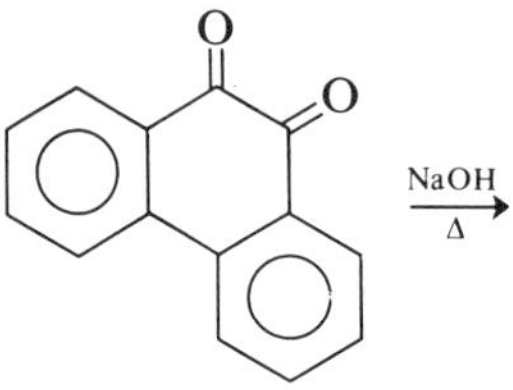

4. (a) Suggest reagents for carrying out the following reaction. (b) Show a mechanism for the reaction. (c) Indicate which aryl group undergoes migration and why.

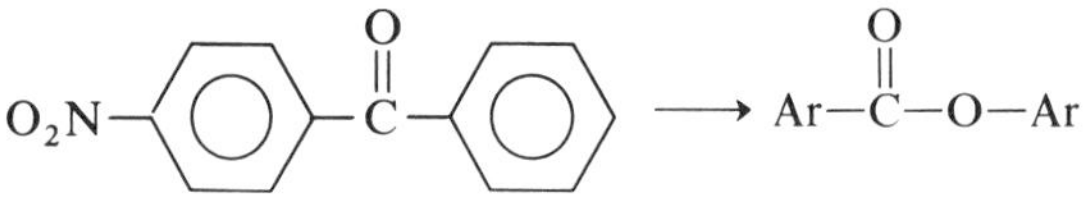

$$O_2N{-}C_6H_4{-}\overset{O}{\overset{\|}{C}}{-}C_6H_5 \longrightarrow Ar{-}\overset{O}{\overset{\|}{C}}{-}O{-}Ar$$

5. Write balanced equations for oxidation or reduction steps in Problems 2 and 3(b).

8

Oxidation and Reduction at Nitrogen, Sulfur, and Phosphorus

The compounds we have treated thus far have usually contained only carbon, hydrogen, and oxygen, and, indeed, the majority of oxidations and reductions of organic compounds involve only those elements. However, many oxidations and reductions of nitrogen-containing groups in organic compounds are known, as well as of groups containing sulfur, phosphorus, and other elements.

FUNCTIONAL GROUPS CONTAINING NITROGEN

The groups containing nitrogen (and carbon-nitrogen bonds) are shown in Table 8-1, arranged roughly in order of decreasing oxidation state. Nitrates ($RONO_2$), nitrites (RONO), and O-substituted hydroxylamines ($RONH_2$) are specifically omitted. Most of the groups shown can be interconverted with others in the list.

Table 8-1

GROUPS CONTAINING NITROGEN IN ORGANIC COMPOUNDS

Group	Name	Oxidation state[a]
$-\overset{\mid}{\underset{\mid}{C}}-NO_2$	nitro	0
$>C=N(\rightarrow O)(OH)$	*aci*-nitro[b]	0
$-C\equiv N\rightarrow O$	nitrile oxide	0

Table 8-1 (*contd.*)

GROUPS CONTAINING NITROGEN IN ORGANIC COMPOUNDS

Group	Name	Oxidation state[a]
$-\overset{\vert}{\underset{\vert}{C}}-NO$	nitroso	2
$>C=N-OH$	oxime	2
$-C\equiv N$	nitrile	2
$-\overset{\vert}{\underset{\vert}{C}}-N=\overset{O\uparrow}{N}-\overset{\vert}{\underset{\vert}{C}}-$	azoxy	3
$-\overset{\vert}{\underset{\vert}{C}}-N_2^+$	diazonium	?[c]
$>C=N_2$	diazo	?[c]
$-\overset{\vert}{\underset{\vert}{C}}-N_3$	azido	?[c]
$-\overset{\vert}{\underset{\vert}{C}}-NHOH$	hydroxylamino	4
$-\overset{\vert}{\underset{\vert}{C}}-N=N-\overset{\vert}{\underset{\vert}{C}}$	azo	4
$>C=NH$	imine	4
$\left(-\overset{\vert}{\underset{\vert}{C}}\right)_3N\rightarrow O$	amine oxide[d]	4
$>C=N-NH_2$	hydrazone	?[c]
$-\overset{\vert}{\underset{\vert}{C}}-NHNH-\overset{\vert}{\underset{\vert}{C}}-$	hydrazo	5
$-\overset{\vert}{\underset{\vert}{C}}-NH_2$	amino	6
$-\overset{\vert}{\underset{\vert}{C}}-NHNH_2$	hydrazino	?[c]

[a] Electrons theoretically required to reduce a nitro group to the oxidation state shown.

[b] The tautomeric form of a nitro group.

[c] Indeterminate; since the second nitrogen is not bound to carbon, it can not be related to a nitro group.

[d] Stable only for tertiary amines.

REDUCTION OF NITRO GROUPS

The nitro group stands at the top of the oxidation scale of groups with nitrogen bonded to carbon; it is in theory capable of being reduced to most of the other groups in Table 8-1. One of the easiest of all reductions to effect is that of a nitro group to an amino group: the reduction of a substituted nitrobenzene to a substituted aniline can be accomplished under the great variety of conditions shown in the following equations. Among useful reducing agents are metals—iron, zinc, tin; salts of metals in their lower oxidation states—ferrous sulfate, stannous chloride; and hydrogen over various catalysts—nickel, platinum. The ease of reduction of the nitro group is underscored by its selective reduction in the presence of aldehyde and ester groups.

CH_3, NO_2, NO_2 $\xrightarrow[C_2H_5OH,\ reflux,\ 2\ hr]{Fe,\ HCl}$ CH_3, NH_2, NH_2

NO_2 $\xrightarrow[C_2H_5OH\ reflux]{Zn}$ 80% NH_2

COOH, O_2N, NO_2, NO_2 $\xrightarrow[100°,\ 1\ hr]{Sn,\ HCl}$ >53% COOH, H_2N, NH_2, NH_2

NO_2, CHO $\xrightarrow[90°]{FeSO_4,\ NH_4OH}$ 72% NH_2, CHO

NO_2, H_3C, CH_3, H_3C, CH_3, NO_2 $\xrightarrow[reflux]{SnCl_2,\ HCl}$ 97% NH_2, H_3C, CH_3, H_3C, CH_3, NH_2

$$m\text{-}O_2NC_6H_4CH(OCH_3)_2 \xrightarrow[CH_3OH,\ 70^\circ]{H_2,\ 70\ \text{atm},\ Ni(R)} m\text{-}H_2NC_6H_4CH(OCH_3)_2$$

$$p\text{-}O_2N\text{—}C_6H_4\text{—}COOC_2H_5 \xrightarrow[Pt,\ r.t.]{H_2,\ 3\ \text{atm}} 95\%\ p\text{-}H_2N\text{—}C_6H_4\text{—}COOC_2H_5$$

Ammonium or sodium sulfide, in the form of the polysulfides Na_2S_X or $(NH_4)_2S_X$, can be used for reducing most aromatic nitro groups to amines, but it is most useful in selectively reducing one of two nitro groups.

$$\text{2,4-}(O_2N)_2C_6H_3OH \xrightarrow[85^\circ]{Na_2S,\ NH_4Cl} 65\%\ \text{2-}H_2N\text{-4-}O_2NC_6H_3OH$$

The reduction of nitro groups to amines is by no means limited to aromatic compounds. As the following equations show, some of the same reducing agents are effective in giving amines from aliphatic nitro compounds, from nitro groups on complex heterocycles, and even from nitrourea and nitroguanidine.

$$(CH_3)_2C(NO_2)(CH_2)_2CO_2C_2H_5 \xrightarrow[Ni,\ C_2H_5OH,\ 50^\circ]{H_2,\ 70\ \text{atm}} >95\%\ (CH_3)_2C(NH_2)(CH_2)_2CO_2C_2H_5$$

$$\text{5-nitrobarbituric acid (5-}NO_2\text{, 5-H)} \xrightarrow[100^\circ]{Sn,\ HCl} 70\%\ \text{5-aminobarbituric acid (5-}NH_2\text{, 5-H)}$$

$$H_2N\text{—}\overset{O}{\overset{\|}{C}}NHNO_2 \xrightarrow[H_2SO_4]{\text{electrolytic}} 65\%\ H_2N\overset{O}{\overset{\|}{C}}NHNH_2$$

nitrourea

$$H_2N\overset{NH}{\overset{\|}{C}}NHNO_2 \xrightarrow[10^\circ]{Zn,\ HOAc} 64\%\ H_2N\text{—}\overset{NH}{\overset{\|}{C}}\text{—}NHNH_2$$

nitroguanidine → aminoguanidine

It is somewhat more difficult to stop the reduction of a nitro group at an intermediate stage. However, under special conditions, usually involving base, it is possible to obtain good yields of compounds with many of the other nitrogen-containing functional groups of Table 8-1. The reductions shown are arranged in descending order of the oxidation states of the products.

$$C_6H_5{-}NO_2 \xrightarrow[\substack{NaOH \\ H_2O,\ \text{reflux},\ 8\ \text{hr}}]{As_2O_3} 85\%\ C_6H_5{-}N{=}\overset{\overset{O}{\uparrow}}{N}{-}C_6H_5$$

azoxybenzene

$$C_6H_5{-}NO_2 \xrightarrow[\substack{NaOH \\ \text{reflux}}]{Zn} 85\%\ C_6H_5{-}N{=}N{-}C_6H_5$$

azobenzene

$$C_6H_5{-}NO_2 \xrightarrow[NH_4Cl]{Zn} 65\%\ C_6H_5{-}NHOH$$

REDUCTION OF OTHER NITROGEN-CONTAINING FUNCTIONAL GROUPS

Like nitro groups, nitroso groups are very readily reduced to the corresponding amines. This can be effected by vigorous reagents such as zinc, but more often the reaction is carried out with a mild reducing agent, especially sodium hydrosulfite ($Na_2S_2O_4$). Not only aromatic but also aliphatic nitroso groups are reduced to amines. N-Nitroso groups, prepared from dialkylamines and nitrous acid, are reduced to unsymmetrical hydrazines by zinc and acetic acid, but more vigorous reagents, such as stannous chloride, give cleavage to the amine with aromatic amines.

$$\text{1-NO-2-ONa-naphthalene} \xrightarrow[35^\circ]{Na_2S_2O_4} 70\%\ \text{1-NH}_2\text{-2-ONa-naphthalene}$$

$$CH_3CO\underset{\displaystyle NO}{\underset{|}{C}}HCO_2C_2H_5 \xrightarrow[\text{reflux}]{Zn,\ HOAc} > 64\%\ CH_3CO\underset{\displaystyle NH_2}{\underset{|}{C}}HCO_2C_2H_5$$

$$\left(\text{actually, the tautomer, } CH_3CO\overset{\overset{\displaystyle NOH}{\|}}{C}CO_2C_2H_5\right)$$

$$(CH_3)_2NH + HONO \xrightarrow[75^\circ]{H_2O} 90\%\ (CH_3)_2N{-}NO$$

$$(CH_3)_2N{-}NO \xrightarrow[60^\circ,\ 1\ hr]{Zn,\ HOAc} 80\%\ (CH_3)_2NNH_2$$

$$m\text{-}CH_3C_6H_4NHC_2H_5 \xrightarrow[10^\circ]{HONO} m\text{-}CH_3C_6H_4N(NO)C_2H_5$$

$$m\text{-}CH_3C_6H_4N(NO)C_2H_5 \xrightarrow[60^\circ]{SnCl_2,\ HCl} m\text{-}CH_3C_6H_4NHC_2H_5$$

Unsymmetrical aromatic azo compounds result from the coupling of diazonium salts with phenols or tertiary aromatic amines. Both they and the symmetrical azo derivatives from the selective reduction of nitro compounds (page 125) behave like nitroso compounds in that they can be reduced to primary amines under relatively mild conditions, e.g., with sodium hydrosulfite. Under very carefully controlled conditions, azo compounds can be reduced to hydrazo compounds, symmetrically disubstituted hydrazines. Monosubstituted hydrazines are produced by reduction of diazonium salts.

$$4\text{-}HO\text{-}C_{10}H_6\text{-}1\text{-}N{=}N{-}C_6H_4\text{-}p\text{-}SO_3Na \xrightarrow[50^\circ]{Na_2S_2O_4} 75\%\ 4\text{-}HO\text{-}C_{10}H_6\text{-}1\text{-}NH_2$$

$$C_6H_5{-}N{=}N{-}C_6H_5 \xrightarrow[CH_3OH]{Zn,\ NaOH} 88\%\ C_6H_5{-}NHNH{-}C_6H_5$$

hydrazobenzene

$$C_6H_5{-}N_2^+Cl^- \xrightarrow[65^\circ]{Na_2SO_3} C_6H_5{-}NHNH_2$$

phenylhydrazine

The two carboxylic acid derivatives that contain nitrogen—amides and nitriles—have already been discussed in Chapter 7 with regard to their utility as intermediates in specialized reductions leading to aldehydes.

More vigorous reduction of either nitriles or amides leads to amines—primary amines from nitriles and primary amides, secondary amines from secondary amides, and tertiary amines from tertiary amides. Lithium aluminum hydride is probably the preferred method on a small scale for reduction of both nitriles and amides, but sodium-ethanol reduction and catalytic hydrogenation, especially over Raney nickel, can also be employed.

$$C_6H_5-CN \xrightarrow{LiAlH_4} C_6H_5-CH_2NH_2$$

$$C_6H_{11}-CON(CH_3)_2 \xrightarrow[\text{reflux, 15 hr}]{LiAlH_4,\ Et_2O,} 88\%\ C_6H_{11}-CH_2N(CH_3)_2$$

$$n\text{-}C_5H_{11}CN \xrightarrow{Na,\ C_2H_5OH} n\text{-}C_5H_{11}CH_2NH_2$$

$$C_6H_5-CH_2CN + \underset{140\text{ atm}}{2H_2} \xrightarrow[130^\circ]{Ni(R)} 85\%\ C_6H_5-CH_2CH_2NH_2$$

Oximes are another group of compounds that yield primary amines on reduction, which can be carried out catalytically or chemically. Since oximes are readily prepared from carbonyl compounds, their reduction constitutes a conversion of aldehydes and ketones to primary amines.

$$n\text{-}C_5H_{11}C(CH_3){=}NOH \xrightarrow[Ni]{H_2,\ 70\text{ atm}} n\text{-}C_5H_{11}CH(CH_3){-}NH_2$$

$$n\text{-}C_6H_{13}CH{=}NOH \xrightarrow[\text{reflux}]{Na,\ C_2H_5OH} 67\%\ n\text{-}C_7H_{15}NH_2$$

Theoretically, one should also be able to prepare primary amines by the hydrogenation of imines ($>C{=}NH$) formed by the addition of ammonia to aldehydes and ketones. Unfortunately, simple imines are not stable, and this is not a useful synthesis. However, imines can be used as intermediates prepared in situ; that is, they are not isolated but are hydrogenated as soon as they are formed. Imines substituted on the nitrogen atom are more stable than simple imines and can be used for the preparation of secondary amines. Sometimes the substituted imines are isolated, but usually they, too, are hydrogenated as formed, in situ. A typical reaction involves heating a mixture of a primary amine, an aldehyde or ketone, and Raney nickel in an atmosphere of hydrogen. Yields are quite high and this is one of the best methods for preparing pure secondary amines. The variety of compounds obtainable is illustrated in the equations.

$$C_6H_5COCH_3 + NH_3 + H_2 \xrightarrow[150^\circ]{Ni(R)} C_6H_5CH(CH_3)NH_2$$
270 atm

$$3\text{-}CH_3C_6H_4NH_2 + C_6H_5CHO + H_2 \xrightarrow{Ni(R)} 92\%\ 3\text{-}CH_3C_6H_4NHCH_2C_6H_5$$
70 atm

$$2,3\text{-}(OCH_3)_2C_6H_3CHO + CH_3NH_2 + H_2 \xrightarrow[C_2H_5OH,\ H_2O,\ r.t.]{Ni(R)} 90\%\ 2,3\text{-}(OCH_3)_2C_6H_3CH_2NHCH_3$$
3 atm

$$(CH_3)_2CO + H_2NCH_2CH_2OH + H_2 \xrightarrow{Pt} 95\%\ (CH_3)_2CHNH(CH_2)_2OH$$
2 atm

One final reduction deserves mention here, the reduction of a quaternary ammonium salt to a tertiary amine and a hydrocarbon by sodium-amalgam. This reduction, called the Emde reduction, has frequently been used in determining the structures of unknown natural products by simplifying the compounds.

$$2,6\text{-}(CH_3)_2C_6H_3CH_2\overset{+}{N}(CH_3)_3I^- \xrightarrow[H_2O,\ 100^\circ,\ 24\ hr]{5\%\ Na(Hg)} 88\%\ 1,2,3\text{-}(CH_3)_3C_6H_3 + (CH_3)_3N$$

OXIDATION OF AMINES

The group containing nitrogen in its lowest oxidation state is clearly the amino group. Since the amino group has farthest to go, one might expect that it should be oxidizable to a variety of other groups of higher oxidation state, depending on the reagent employed. This is certainly true: one can oxidize a primary amino group all the way to a nitro group or stop at intermediate stages, depending on the conditions employed. To oxidize an aromatic amino group to a nitro group, a powerful peracid, especially potassium persulfate in sulfuric acid (Caro's acid) or pertrifluoroacetic acid, is employed, as illustrated for aniline and *o*-nitroaniline. Under carefully controlled conditions, oxidation of an aromatic primary amine to a nitroso compound is also possible with Caro's acid.

$$C_6H_5NH_2 \xrightarrow{HOSO_2OH(O)} C_6H_5NO_2$$

$$o\text{-}NO_2C_6H_4NH_2 \xrightarrow{CF_3CO_3H} o\text{-}C_6H_4(NO_2)_2$$

$$2\text{-}CH_3\text{-}5\text{-}O_2N\text{-}C_6H_3NH_2 \xrightarrow[H_2SO_4,\ 40^\circ]{K_2S_2O_8} 63\%\ 2\text{-}CH_3\text{-}5\text{-}O_2N\text{-}C_6H_3NO$$

Tertiary amines behave differently towards peracids, yielding amine oxides. This conversion can be effected with hydrogen peroxide alone, with Caro's acid, or with other peracids.

$$R_3N \xrightarrow[\text{or } C_6H_5CO_3H \text{ or } H_2O_2 \text{ in } RCOOH]{HSO_3O_2H \text{ or } CH_3CO_3H} R_3\overset{+}{N}-\overset{-}{O}$$

$$C_6H_{11}CH_2N(CH_3)_2 \xrightarrow[\text{r.t., 36 hr}]{15\%H_2O_2} C_6H_{11}CH_2\overset{+}{N}(O^-)(CH_3)_2$$

A very useful oxidation of aromatic amines is their well-known conversion by nitrous acid to diazonium salts, intermediates employed for the synthesis of a wide variety of aromatic compounds. The syntheses of diazonium salts and their many reactions are discussed in detail elsewhere in this series.* As we saw on page 126, secondary amines react with nitrous acid to give N-nitroso compounds, the starting materials for the preparation of mono-substituted hydrazines.

$$m\text{-}H_2NC_6H_4CHO \xrightarrow[HCl]{NaNO_2} >60\%\ m\text{-}OHCC_6H_4N_2^+Cl^-$$

* L. M. Stock, *Aromatic Substitution Reactions*, in Foundations of Modern Organic Chemistry Series, Prentice-Hall, Inc., Englewood Cliffs, N. J., 1968, pp. 92–100.

Another important reaction is the oxidation of aromatic amines with *o*- or *p*-hydroxyl or amino groups to *o*- or *p*-quinones. This reaction, like the oxidation of phenols, is much used in making dyes. Oxidation can be carried out with either mild or vigorous oxidizing agents—from ferric chloride to potassium dichromate, as shown in the examples. The intermediate imines are usually hydrolyzed to the carbonyl groups of the quinones but can sometimes be isolated, as shown in the third example. Since they are so susceptible to oxidation, aromatic amines are often stored in the presence of inhibitors to prevent autoxidation.

$FeCl_3$, 30° → 90%

$NaNO_2$, H_2SO_4, 60° → >80%

NaOCl(NaOH, Cl_2), 15° → 85%

dil HNO_3 → 95%

$K_2Cr_2O_7$, H_2SO_4, 90° → 80%

OXIDATION OF OTHER NITROGEN-CONTAINING GROUPS

Oxidations of other nitrogen-containing groups often require controlled conditions. For example, the conversion of a hydroxylamine to a nitroso group requires strict temperature control, since the nitroso group itself is susceptible to further oxidation. This is illustrated in the two reactions shown, where the nitroso group is both a product and a reactant in dichromate. The preparation of nitrosobenzene from hydroxylaminobenzene, when combined with the reduction of nitrobenzene to hydroxylaminobenzene (page 125), constitutes, in effect, a reduction of nitrobenzene to nitrosobenzene, a reaction that cannot be carried out directly.

$$C_6H_5-NHOH \xrightarrow[H_2SO_4,\ -5^\circ]{Na_2Cr_2O_7} 50\%\ C_6H_5-NO$$

nitrosobenzene

$$4\text{-}O_2N\text{-}2\text{-}(NO)C_6H_3CH_3 \xrightarrow[H_2SO_4,\ 65^\circ]{K_2Cr_2O_7} 60\%\ 4\text{-}O_2N\text{-}2\text{-}(NO_2)C_6H_3COOH$$

Hydrazo compounds are very unstable to oxidation and are converted readily to azo compounds. The latter are more stable but give azoxy compounds with peroxides.

$$C_6H_{10}(CN)-NHNH-(NC)C_6H_{10} \xrightarrow[15^\circ]{Br_2} 87\%\ C_6H_{10}(CN)-N{=}N-(NC)C_6H_{10}$$

$$C_2H_5O_2C-NHNH-CO_2C_2H_5 \xrightarrow[5^\circ]{\text{concd } HNO_3} C_2H_5O_2C-N{=}N-CO_2C_2H_5$$

$$C_6H_5-N{=}N-C_6H_5 \xrightarrow[H^+]{H_2O_2} C_6H_5-\overset{O}{\overset{\uparrow}{N}}{=}N-C_6H_5$$

A special oxidation leading to useful intermediates is the oxidation of hydrazones to diazo compounds with the mild oxidant mercuric oxide. The diazo compounds lose nitrogen when heated or irradiated. The resulting carbenes dimerize, as illustrated in the formation of tetraphenylethene, or rearrange, as illustrated for the formation of diphenylketene.

$$(C_6H_5)_2C{=}NNH_2 \xrightarrow[\text{r.t.}]{\text{HgO}} 93\%\ (C_6H_5)_2CN_2 + Hg + H_2O$$

$$(C_6H_5)_2CN_2 \xrightarrow{\Delta} (C_6H_5)_2C{=}C(C_6H_5)_2 + 2N_2$$

$$C_6H_5C({=}NNH_2)C({=}O)C_6H_5 \xrightarrow{\text{HgO}} C_6H_5C(N_2)C({=}O)C_6H_5 \xrightarrow{100^\circ} 60\%\ (C_6H_5)_2C{=}C{=}O + N_2$$

SULFUR-CONTAINING FUNCTIONAL GROUPS

In the first section of this chapter, nitrogen-containing groups were arranged according to their oxidation states. A similar list of organic sulfur compounds containing carbon-sulfur bonds, arranged in decreasing order of oxidation state, is shown in Table 8-2. A striking feature of Table 8-2 is the very large number of groups at the lowest oxidation state. These are all in theory, and nearly all in practice, derivable from hydrogen sulfide. Since we are discussing oxidation and reduction and most of the reactions of those groups do not normally involve oxidation or reduction, we shall direct our attention to the other groups. We shall also restrict our attention to those compounds containing carbon-sulfur bonds, omitting sulfates, $ROSO_2OH$ and $ROSO_2OR$, and sulfites, ROSOOH and ROSOOR. In contrast to the nitrogen-containing groups, where a majority were derivable by reduction of nitro compounds, the highest oxidation state, the major route to most of the groups in Table 8-2 is through oxidation of mercaptans, sulfides, and compounds of the lowest oxidation state. We turn first, then, to oxidation.

OXIDATION OF MERCAPTANS, SULFIDES, AND OTHER GROUPS

The oxidation products obtainable from mercaptans are quite different from those obtainable from sulfides, just as the oxidation products of primary and secondary amines often differ. The easiest oxidation of mercaptans is

Table 8-2

SULFUR-CONTAINING ORGANIC GROUPS

Group	Name	Oxidation state[a]
$R-S(=O)_2-OH$	sulfonic acid	0
$R-S(=O)-OH$	sulfinic acid	2
$R-S(=O)_2-R$	sulfone	2
$R-S-OH$	sulfenic acid	4
$R-S(=O)-R$	sulfoxide	4
$R-S-S-R$	disulfide	5
$R-SH$	mercaptan, thiol	6
$R-S-R$	sulfide, thioether	6
$R-\overset{+}{S}(R)-R$	sulfonium salt	6
$R-C(=S)-H$	thioaldehyde	6
$R-C(=S)-R$	thioketone	6
$R-C(=S)-OH$	thion acid	6
$R-C(=O)-SH$	thiol acid	6
$R-C(=S)-SH$	dithio acid	6

[a] Electrons theoretically required to reduce a sulfonic acid group to the oxidation state shown.

their conversion to disulfides. This can be carried out with quite mild reagents, among them halogens, ferric ion, and air. Thiol acids behave in the same way.

$$n\text{-}C_5H_{11}SH \xrightarrow{Cl_2} n\text{-}C_5H_{11}-S-S-C_5H_{11}$$

$$C_6H_5SH \xrightarrow{O_2 \text{ or } Fe^{3+} \text{ or } I_2} C_6H_5-S-S-C_6H_5$$

$$H_2N-C_6H_4-SNa \xrightarrow[70^\circ]{3\%\ H_2O_2} >60\%\ H_2N-C_6H_4-S-S-C_6H_4-NH_2$$

$$C_6H_5-CO-SK \xrightarrow{I_2} >70\%\ C_6H_5-CO-S-S-CO-C_6H_5$$

benzoyl disulfide

The disulfides formed are unstable in the presence of excess halogen, however, and are sometimes oxidized further to sulfenyl chlorides or even to sulfonic acids and their derivatives.

$$(n\text{-}C_5H_{11}S)_2 \xrightarrow{Cl_2} 78\%\ n\text{-}C_5H_{11}SCl$$

1-pentanesulfenyl chloride

$$(o\text{-}NO_2C_6H_4S-)_2 \xrightarrow[HNO_3,\ 70^\circ]{Cl_2} 84\%\ o\text{-}NO_2C_6H_4SO_2Cl$$

$$(HOOC-\underset{NH_2}{CH}-CH_2S)_2 \xrightarrow[aq\ HCl,\ 60^\circ]{Br_2} HOOC\underset{NH_2}{C}HCH_2SO_3H$$

cystine — cysteic acid

Sulfides are also susceptible to oxidation. They react with halogen to give adducts that are hydrolyzed to sulfoxides. Sulfoxides are formed from sulfides by a number of other reagents as well—periodate, peracids, etc. Further oxidation of the sulfoxides with peroxides or peracids gives sulfones. More vigorous oxidizing agents, such as nitric acid, can give sulfonic acids.

$$R_2S \xrightarrow{Br_2} R_2SBr_2 \xrightarrow{H_2O} R_2SO$$

$$R_2S \xrightarrow{NaIO_4} R_2SO$$

$$R_2S \xrightarrow{C_6H_5CO_3H\text{ or }CH_3CO_3H,\text{ etc.}} R_2SO$$

$$R_2SO \xrightarrow{HSO_3O_2H} R_2SO_2$$

$$\text{tetrahydrothiophene} \xrightarrow{30\%\ H_2O_2} \text{tetrahydrothiophene }S{=}O \xrightarrow{30\%\ H_2O_2} \text{tetrahydrothiophene }O{=}S{=}O$$

$$(n\text{-}C_4H_9)_2S \xrightarrow{HNO_3} n\text{-}C_4H_9SO_3H$$

REDUCTION OF SULFUR-CONTAINING GROUPS

Sulfonic acids are quite resistant to reduction. However, sulfonyl chlorides, their acid chlorides, can be reduced easily to sulfinic acids or, under more vigorous conditions, all the way to mercaptans.

$$CH_3CONH-C_6H_4-SO_2Cl \xrightarrow[\text{r.t.}]{Na_2SO_3} 45\%\ CH_3CONH-C_6H_4-SO_2H$$

$$H_3C-C_6H_4-SO_2Cl \xrightarrow[(2)\ Na_2CO_3]{(1)\ Zn,\ 80^\circ} 64\%\ H_3C-C_6H_4-SO_2Na$$

$$\text{1,5-}C_{10}H_6(SO_2Cl)_2 \xrightarrow[\text{reflux}]{Zn(Hg),\ 12N\ H_2SO_4} \text{1,5-}C_{10}H_6(SH)_2$$

Disulfides and polysulfides can also easily be reduced to mercaptans.

$$(o\text{-}HOOC-C_6H_4-S-)_2 \xrightarrow[\text{reflux}]{Zn,\ HOAc} 80\%\ o\text{-}HOOC-C_6H_4-SH$$

$$(C_2H_5O)_2CHCH_2-S_n-CH_2CH(OC_2H_5)_2 \xrightarrow[\text{liq } NH_3]{Na} > 77\%\ NaSCH_2CH(OC_2H_5)$$

We saw in Chapter 2 that nickel is an effective scavenger of sulfur and in Chapter 4 that this property can be utilized in an effective conversion of ketones and aldehydes, via their thioketals and thioacetals, to hydrocarbons. An additional example is shown here.

$$(CH_2)_9\,C(S-CH_2)_2 \text{ (cyclic)} \xrightarrow{Ni} 80\%\ (CH_2)_{10} \text{ (cyclic)} + 2H_2S + C_2H_6$$

Nickel can be generally used to remove sulfur, as long as the sulfur removed is in its lowest oxidation state, as is the case for one of the sulfurs in the alkyl thiosulfate shown. The latter compound would be formed by treatment of a halide with thiosulfate ($S_2O_3^{2-}$) ion.

$$C_6H_5-CH_2CH_2SSO_2^-Na^+ \xrightarrow[NaOD]{Ni(Al)} C_6H_5-CH_2CH_2D$$

OXIDATION AND REDUCTION OF NITROGEN AND SULFUR HETEROCYCLES

We have treated saturated heterocyclic compounds as representatives of secondary or tertiary amines or sulfides in their preceding reactions, as indeed they are. For unsaturated heterocycles like the pyridine ring, the categorization is blurred, but the pyridine ring, too, is a tertiary amine and behaves like one, giving an N-oxide with hydrogen peroxide, for example.

$$\xrightarrow[100^\circ]{5\%\ H_2O_2,\ HOAc} 78\%$$

CONH$_2$ ... N ... CONH$_2$ ... N$^+$–O$^-$

There are, however, other reactions more peculiar to the heterocycles. Two of these are the reactions of saturation and desaturation. In Chapter 2, we saw that heterocyclic rings behave like other potentially aromatic systems; thus, the usual hydrogenation and dehydrogenation techniques can be applied to them. Other examples are shown here. Pyrrolines give pyrroles on selenium dehydrogenation and a dihydropyridine is very easily oxidized, as expected. Chemical reduction of pyridine is somewhat easier than the corresponding reduction of benzene.

$$\xrightarrow[250^\circ]{Se} 45\%$$

ϕ ... N H ... ϕ ... ϕ ... N H ... ϕ

$C_2H_5O_2C$ ... $CO_2C_2H_5$ ... H_3C ... N H ... CH_3 $\xrightarrow[\text{dil } H_2SO_4,\ 50^\circ]{\text{dil } HNO_3}$ $C_2H_5O_2C$ ... $CO_2C_2H_5$ ... H_3C ... N ... CH_3

$$\xrightarrow[C_2H_5OH,\ \text{reflux}]{Na} 75\%$$

N ... N H

Heterocyclic ring systems can also be destroyed by vigorous oxidation, just as homocyclic aromatic systems can. We can draw conclusions regarding the relative stabilities of ring systems from their ease of oxidation or reduction. The pyridine ring, for example, is remarkably stable toward oxidation in acid. However, the whole of heterocyclic chemistry is a somewhat complex and specialized subject, which will be dealt with elsewhere in the Foundations of Modern Organic Chemistry Series.*

conc HNO$_3$, 95°, 12 hr → 70%

KClO$_3$, HCl → >62%

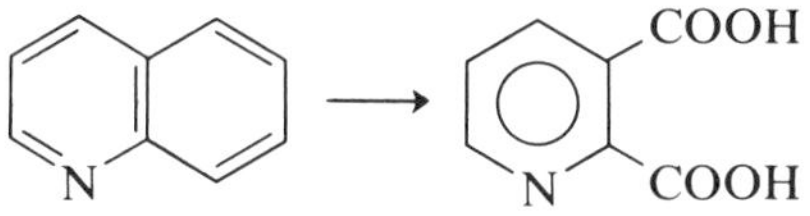

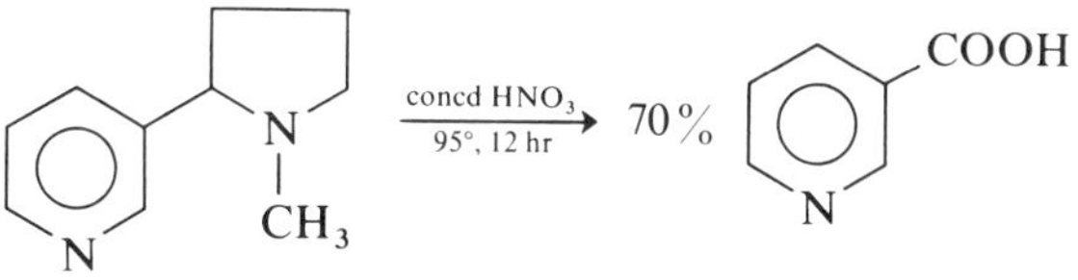

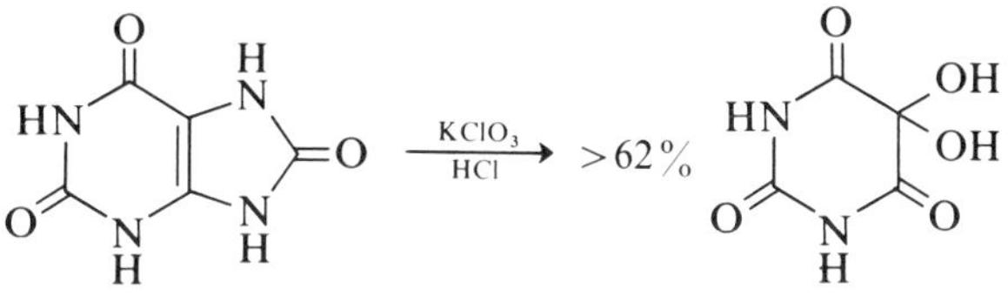

uric acid

alloxan hydrate

PHOSPHORUS-CONTAINING GROUPS

Heteroatoms other than oxygen, nitrogen, or sulfur are less common in organic chemistry. The next most important element is probably phosphorus, which is found in reagents useful for aldehyde and ketone condensations in nucleosides, nucleic acids and related compounds,† and elsewhere.

Phosphorus stands just below nitrogen in the periodic table, and we would expect it to have similar oxidation states. The more common groups containing carbon-phosphorus bonds are shown in Table 8-3. Phosphates, $(RO)_3PO$, and phosphites, $(RO)_3P$, are omitted because they do not contain carbon-phosphorus bonds. Much of the chemistry of the compounds in Table 8-3 depends on the number of hydrogen atoms (as opposed to alkyl or aryl groups) attached to the phosphorus. For example, phosphine (PH_3) is explosive but trialkyl or triarylphosphines are quite stable. Some pairs of compounds exist at the same oxidation state: phosphonous acids and phosphinic acids, phosphinous acids and phosphine oxides. In these pairs, the structures depend on the number of alkyl (or aryl) groups attached to phosphorus.

* E. C. Taylor, *Heterocyclic Chemistry*, Prentice-Hall, Inc., Englewood Cliffs, N. J., in preparation.

† Robert Barker, *Organic Chemistry of Biological Compounds*, in Foundations of Modern Organic Chemistry Series, Prentice-Hall, Inc., Englewood Cliffs, N. J., 1971, Chapters 6 and 8.

Table 8-3

GROUPS CONTAINING PHOSPHORUS IN ORGANIC COMPOUNDS

Group	Name	Oxidation level[a]
$RP{=}O(OH)_2$	phosphonic acid	0
$RP(OH)_2$	phosphonous acid	2
$R_2P{=}O(OH)$	phosphinic acid	2
R_2POH	phosphinous acid	4
R_3PO	phosphine oxide	4
PH_3, PR_3	phosphine	6
R_5P or $R_3P{=}CR_2$	phosphorane	8

[a] Electrons theoretically required to reduce a phosphonic acid group to the oxidation state shown.

OXIDATIONS AND REDUCTIONS INVOLVING PHOSPHORUS-CONTAINING GROUPS

Perhaps the most striking aspect of organophosphorus chemistry is the great ease of oxidation of phosphines. These compounds, like amines, are nucleophiles, readily attacked by oxidizing species; peroxides are especially effective. An interesting mechanistic study using ^{18}O-labelled dibenzoyl peroxide has demonstrated that it is the singly bonded oxygen of the peracid that provides the oxygen of the phosphine oxide. Other peroxides also serve as oxidants, as illustrated for cumene hydroperoxide.

$$(C_6H_5)_3P + (C_6H_5C({=}^{18}O){-}O)_2 \longrightarrow (C_6H_5)_3P{=}O + (C_6H_5C({=}O))_2O$$

(containing two ^{18}O atoms)

$$[CH_3(CH_2)_3]_3P + C_6H_5C(CH_3)_2{-}O{-}OH \longrightarrow (C_4H_9)_3P{=}O + C_6H_5C(CH_3)_2{-}OH$$

Oxygen can also convert phosphines to phosphine oxides, but the reaction also gives phosphinates as by-products.

$$2[CH_3(CH_2)_3]_3P + \tfrac{3}{2}O_2 \xrightarrow[\text{benzene}]{0^\circ} (C_4H_9)_3P{=}O + (C_4H_9)_2P({=}O)OC_4H_9$$

Phosphines are so readily oxidized that they can deoxygenate a wide variety of other organic compounds and serve as rather effective and selective reducing agents. This is illustrated in the conversions shown—of an epoxide to an alkene, of azoxybenzene to azobenzene, and of a sulfoxide to a sulfide.

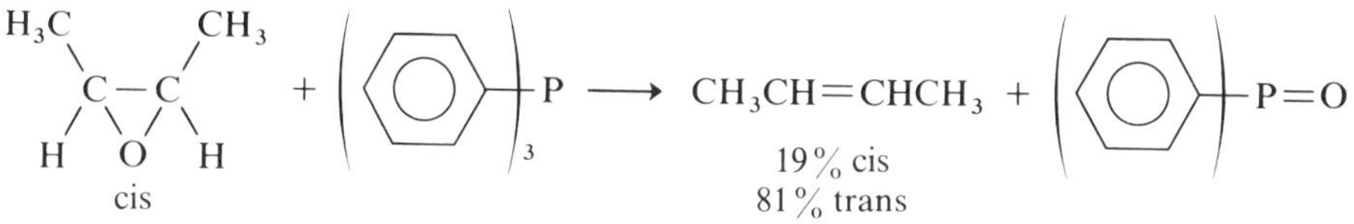

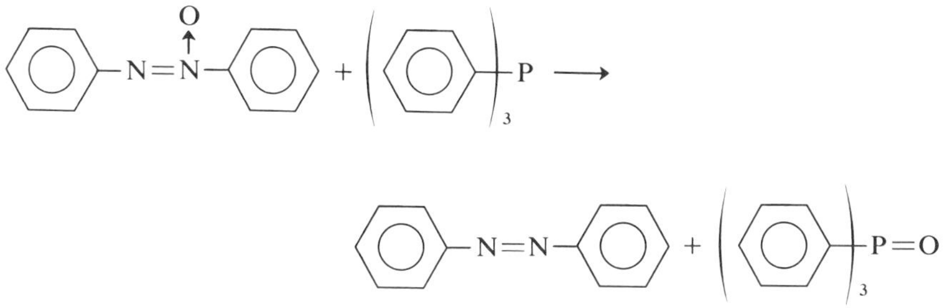

$$\text{cis-2,3-epoxybutane} + (C_6H_5)_3P \longrightarrow CH_3CH{=}CHCH_3 + (C_6H_5)_3P{=}O$$

cis; 19% cis, 81% trans

$$C_6H_5{-}N{=}N(\rightarrow O){-}C_6H_5 + (C_6H_5)_3P \longrightarrow C_6H_5{-}N{=}N{-}C_6H_5 + (C_6H_5)_3P{=}O$$

$$R_2S{=}O + [CH_3(CH_2)_3]_3P \longrightarrow R_2S + (C_4H_9)_3P{=}O$$

Many of the reactions just described for phosphines are also applicable to phosphites, which also can serve as effective deoxygenating agents. Phosphites can be useful in desulfurizing mercaptans. They undergo oxidation to phosphates by peroxides.

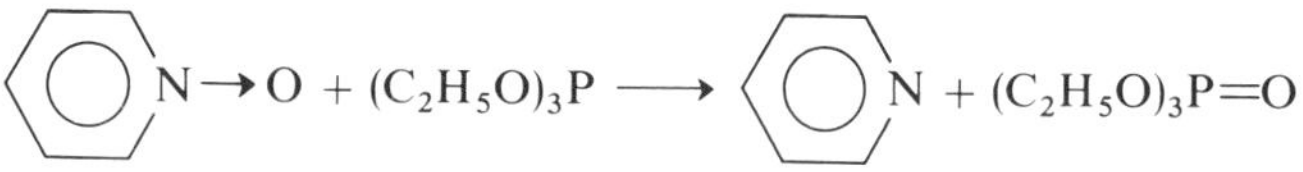

$$C_5H_5N{\rightarrow}O + (C_2H_5O)_3P \longrightarrow C_5H_5N + (C_2H_5O)_3P{=}O$$

$$RSH + (RO)_3P \longrightarrow RH + (RO)_3P{=}S$$

$$(CH_3)_3CO{-}OH + (CH_3CH_2O)_3P \longrightarrow (CH_3)_3COH + (C_2H_5O)_3P{=}O$$

Finally, we should note here some reactions of phosphonium salts, derived by treatment of a tertiary phosphine with an alkyl halide. The most important reaction of phosphonium salts is their conversion by base to phosphoranes, also called phosphorylidenes. The latter compounds serve as intermediates in the Wittig reaction, which is discussed elsewhere in the Foundations of Modern Organic Chemistry Series.* The Wittig reaction is illustrated here for the preparation of methylenecyclohexane.

* C. D. Gutsche, *The Chemistry of Carbonyl Compounds*, Prentice-Hall, Inc., Englewood Cliffs, N. J., 1967, pp. 72–73.

$$(C_6H_5)_3P + CH_3I \longrightarrow (C_6H_5)_3\overset{+}{P}{-}CH_3\,I^- \longrightarrow (C_6H_5)_3P{=}CH_2$$

methyltriphenylphosphonium iodide — triphenylmethylene-phosphorane

$$(C_6H_5)_3P{=}CH_2 + O{=}C_6H_{10} \longrightarrow (C_6H_5)_3P{=}O + H_2C{=}C_6H_{10}$$

A Wittig reaction represents an oxidation of the phosphorus atom. The intermediate phosphoranes can also be oxidized by other reagents to phosphine oxides.

$$(C_6H_5)_3P{=}CH_2 + \begin{matrix} O_2 \\ \text{or} \\ CH_3CO_3H \end{matrix} \longrightarrow (C_6H_5)_3P{=}O + CH_2O$$

Phosphonium salts can be reduced by lithium aluminum hydride to tertiary phosphines and hydrocarbons. This reaction is analogous to the Emde reduction of quaternary ammonium salts.

$$(C_6H_5)_3\overset{+}{P}{-}CH_3 + LiAlH_4 \longrightarrow (C_6H_5)_2PCH_3 + C_6H_6$$

OXIDATIONS AND REDUCTIONS AT OTHER ELEMENTS

Immediately below phosphorus in the periodic table (and two rows below nitrogen) is arsenic. Although arsenic is much more metallic than nitrogen or phosphorus, it undergoes some of the same reactions. For example, an arsine can be converted to an arsine oxide, and an arsenic analogue of an azo compound can be prepared.

$$\underset{\text{triphenylarsine}}{(C_6H_5)_3As} + 6\%\ H_2O_2 \xrightarrow[30^\circ]{} 85\%\ \underset{\text{triphenylarsine oxide}}{(C_6H_5)_3AsO}$$

$$NaO{-}\overset{\overset{O}{\|}}{\underset{\underset{ONa}{|}}{As}}{-}CH_2CO_2Na \xrightarrow[\text{r.t.}]{H_3PO_2} 74\%\ \begin{matrix} AsCH_2CO_2H \\ \| \\ AsCH_2CO_2H \end{matrix}$$

Oxidation states of halogens higher than the usual halide level are not uncommon and can be illustrated in the following reaction sequence. Like nitroso compounds, the iodoso compound is rather unstable, being easily oxidized to the iodoxy derivative and disproportionating to it and the iodide on heating.

C_6H_5–I $\xrightarrow[CHCl_3,\ -20°]{Cl_2}$ 90% C_6H_5–ICl_2 $\xrightarrow[70°]{NaOCl}$ 90% C_6H_5–IO_2

iodobenzene dichloride — iodoxybenzene 95%

C_6H_5–ICl_2 $\xrightarrow[NaOH,\ r.t.,\ overnight]{Na_2CO_3}$ C_6H_5–IO (iodosobenzene)

C_6H_5–IO $\xrightarrow{steam\ distill}$ C_6H_5–I + C_6H_5–IO_2

REFERENCES

Augustine (*Oxidation*): pp. 53–118, D. G. Lee, "Oxidation of Oxygen- and Nitrogen-Containing Functional Groups with Transition Metal Compounds."

Hudson, R. F., *Structure and Mechanism in Organo-Phosphorus Chemistry*, Academic Press, Inc., New York, 1965.

Kharasch, N., ed., *Organic Sulfur Compounds*, Vol. 1, Pergamon Press, Oxford, 1961.

Kharasch, N., and C. Y. Meyers, eds., *Organic Sulfur Compounds*, Vol. 2, Pergamon Press, Oxford, 1966.

Kirby, A. J., and S. G. Warren, *Organic Chemistry of Phosphorus*, Elsevier Publishing Co., Amsterdam, 1967.

Millar, I. T., and H. D. Springall, *A Shorter Sidgwick's Organic Chemistry of Nitrogen*, Clarendon Press, Oxford, 1969.

Rylander, P. N., *Catalytic Hydrogenation over Platinum Metals*, Academic Press, Inc., New York, 1967, Part III ("Hydrogenation of Nitrogen Compounds"). *Ibid.*, Chapter 16 ("Reductive Alkylation").

Sasse, K., "Organophosphorus Compounds," in *Methoden der Organischen Chemie* (*Houben-Weyl*), Vol. 12/1 and 12/2, E. Müller, ed., Georg Thieme Verlag, Stuttgart, 1963 and 1964.

Schröter, R., "Amines by Reduction," in *Methoden der Organischen Chemie* (*Houben-Weyl*), Vol. 2/1, E. Müller, ed., Georg Thieme Verlag, Stuttgart, 1957, Chap 4.

Sidgwick, N. V., *The Organic Chemistry of Nitrogen*, 3rd ed., Clarendon Press, Oxford, 1966.

Smith, P. A. S., *The Chemistry of Open-Chain Nitrogen Compounds*, Vols. 1 and 2, W. A. Benjamin, Inc., New York, 1966.

Waters: Chap. 9, "Oxidation of Phenols and Aromatic Amines."

In addition, a number of references found in Chapter 2 also deal with nitrogen-containing compounds.

PROBLEMS

1. Write a detailed mechanism for each of the following reactions, including all intermediates and/or transition states.

(a) $C_6H_5-NH_2 \xrightarrow[HCl]{NaNO_2} C_6H_5-N_2^+Cl^-$

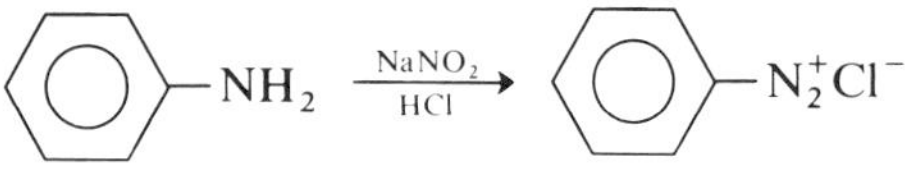

(b) $CH_3\overset{O}{\overset{\|}{C}}CH_2CH_2COOH + H_2 \xrightarrow[Ni(R),\ 140°]{CH_3NH_2}$ (100 atm) 5-methyl-1-methyl-2-pyrrolidinone (H_3C, N–CH_3, =O)

(c) o-CH_3O–C_6H_4–$CH{=}C(CH_3)NO_2 \xrightarrow[5N\ HCl,\ 75°]{Fe,\ FeCl_3}$ o-CH_3O–C_6H_4–CH_2COCH_3

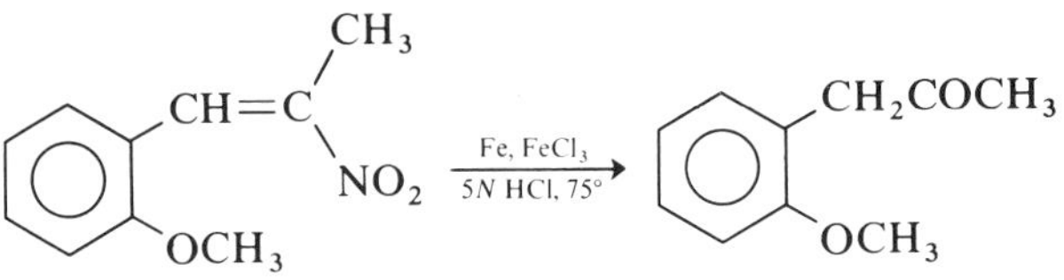

2. Suggest reaction conditions (reagents, catalysts, solvents, etc.) suitable for effecting the following conversions. More than one step may be required.

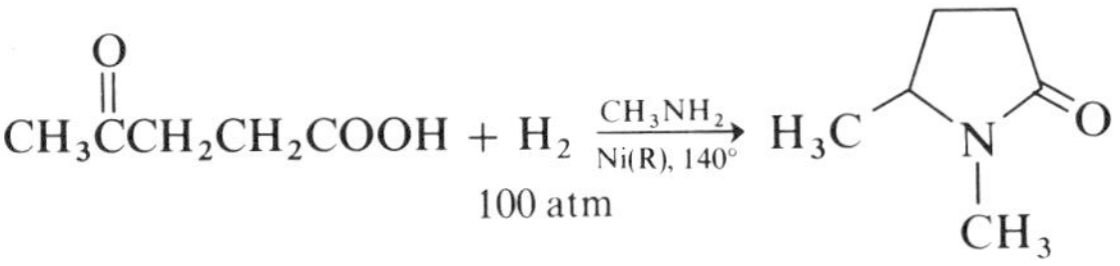

(a) o-NH_2–C_6H_4–COOH $\longrightarrow$ o-$\overset{+}{N}_2$–C_6H_4–COOH

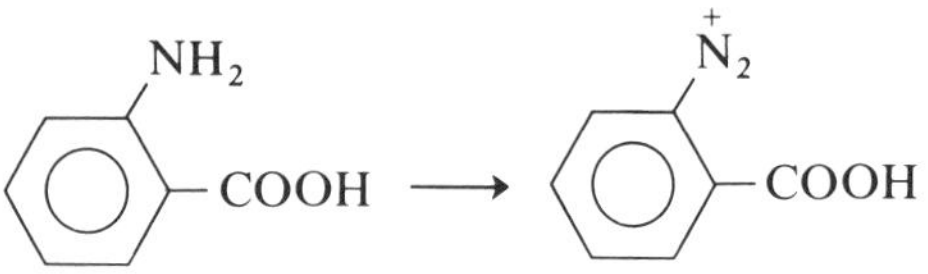

(b) pyridine $\longrightarrow$ pyridine N-oxide (N→O)

(c) $C_6H_5-SO_2Cl \longrightarrow C_6H_5-SH$

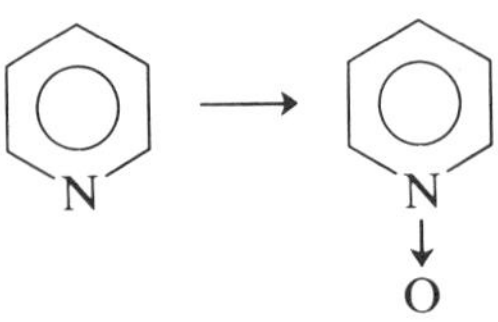

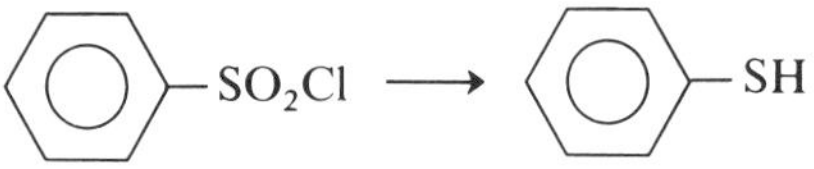

(d) NH_2, Cl, OH (benzene ring) $\longrightarrow$ O, Cl, O (ring)

(e) $C_2H_5CH(NO_2)CH_3 \longrightarrow C_2H_5CH(NH_2)CH_3$

(f) $N{\equiv}C(CH_2)_8C{\equiv}N \longrightarrow H_2N(CH_2)_{10}NH_2$

(g) $C_6H_5\text{—}NHCH_3 \longrightarrow C_6H_5\text{—}N(NH_2)\text{—}CH_3$

(h) CHO, NO_2 (benzene ring) $\longrightarrow$ CHO, NH_2 (benzene ring)

(i) NH_2, NO_2, NO_2 (benzene ring) $\longrightarrow$ NH_2, NH_2, NO_2 (benzene ring)

(j) $O_2N\text{—}C_6H_5 \longrightarrow HONH\text{—}C_6H_5$

(k) $O_2N\text{—}C_6H_5 \longrightarrow C_6H_5\text{—}N(\rightarrow O){=}N\text{—}C_6H_5$

(l) $O_2N\text{—}C_6H_5 \longrightarrow C_6H_5\text{—}N{=}N\text{—}C_6H_5$

3. Indicate all of the expected products of the following reaction.

(a) 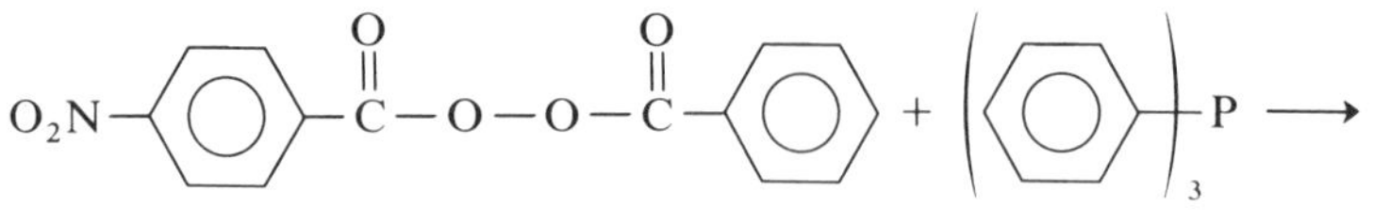 $O_2N-C_6H_4-C(=O)-O-O-C(=O)-C_6H_5 + (C_6H_5)_3P \longrightarrow$

(b) 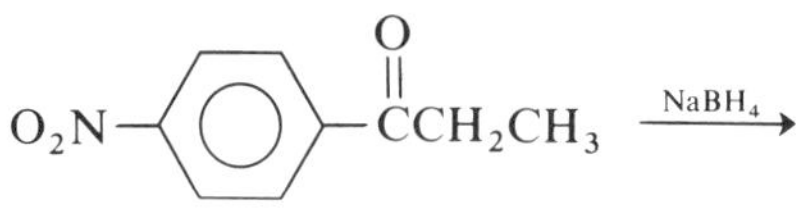 $O_2N-C_6H_4-C(=O)CH_2CH_3 \xrightarrow{NaBH_4}$

4. Write balanced equations for the oxidation or reduction steps in Problems 1(a), 1(b), 1(c), 2(c), 2(d), 2(h), 2(j), and 2(l).

5. Balance the following equation. What are the oxidation levels of the two nitrogens in the products?

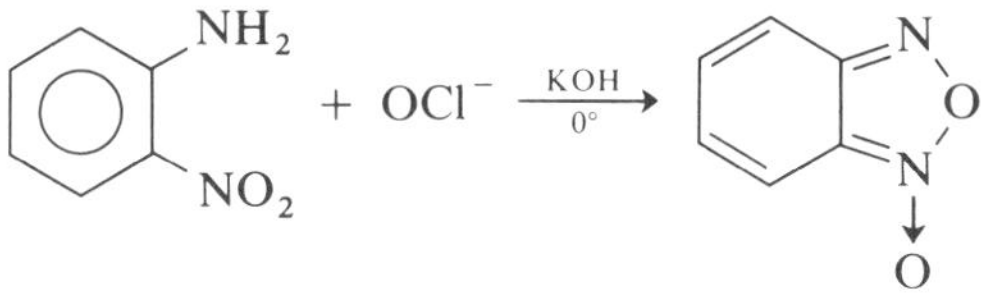

$o\text{-}NH_2C_6H_4NO_2 + OCl^- \xrightarrow[0^\circ]{KOH}$ benzofuroxan (N, O, N→O)

6. What is the reducing agent in the following reaction?

3-$CONH_2$-pyridine N→O $\xrightarrow[POCl_3]{PCl_5}$ 3-CN-2-Cl-pyridine (N)

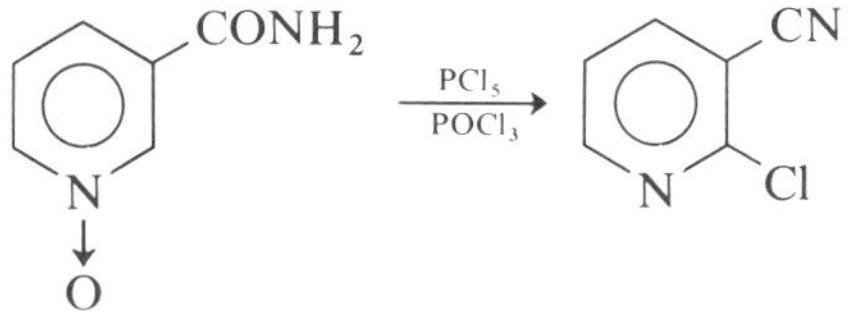

7. In Table 8-2 on page 133, what would be the hypothetical oxidation states of the following sulfur-containing groups: RSO_2Cl (sulfonyl chloride), $RSCl$ (sulfenyl chloride), $R_2C(SR')_2$ (thioketal)?

Index

M